Shahin Kazeminejad

Impact of Solar Particle Events on the Earth's Atmosphere

Shahin Kazeminejad

Impact of Solar Particle Events on the Earth's Atmosphere

Investigations of the Mesosphere and Stratosphere over a complete Solar Cycle

Südwestdeutscher Verlag für Hochschulschriften

Impressum/Imprint (nur für Deutschland/ only for Germany)
Bibliografische Information der Deutschen Nationalbibliothek: Die Deutsche Nationalbibliothek
verzeichnet diese Publikation in der Deutschen Nationalbibliografie; detaillierte bibliografische
Daten sind im Internet über http://dnb.d-nb.de abrufbar.
Alle in diesem Buch genannten Marken und Produktnamen unterliegen warenzeichen-, marken-
oder patentrechtlichem Schutz bzw. sind Warenzeichen oder eingetragene Warenzeichen der
jeweiligen Inhaber. Die Wiedergabe von Marken, Produktnamen, Gebrauchsnamen,
Handelsnamen, Warenbezeichnungen u.s.w. in diesem Werk berechtigt auch ohne besondere
Kennzeichnung nicht zu der Annahme, dass solche Namen im Sinne der Warenzeichen- und
Markenschutzgesetzgebung als frei zu betrachten wären und daher von jedermann benutzt
werden dürften.

Verlag: Südwestdeutscher Verlag für Hochschulschriften Aktiengesellschaft & Co. KG
Dudweiler Landstr. 99, 66123 Saarbrücken, Deutschland
Telefon +49 681 37 20 271-1, Telefax +49 681 37 20 271-0, Email: info@svh-verlag.de
Zugl.: Bremen, University of Bremen, Dissertation, 2009

Herstellung in Deutschland:
Schaltungsdienst Lange o.H.G., Berlin
Books on Demand GmbH, Norderstedt
Reha GmbH, Saarbrücken
Amazon Distribution GmbH, Leipzig
ISBN: 978-3-8381-0902-2

Imprint (only for USA, GB)
Bibliographic information published by the Deutsche Nationalbibliothek: The Deutsche
Nationalbibliothek lists this publication in the Deutsche Nationalbibliografie; detailed
bibliographic data are available in the Internet at http://dnb.d-nb.de.
Any brand names and product names mentioned in this book are subject to trademark, brand or
patent protection and are trademarks or registered trademarks of their respective holders. The
use of brand names, product names, common names, trade names, product descriptions etc.
even without a particular marking in this works is in no way to be construed to mean that such
names may be regarded as unrestricted in respect of trademark and brand protection legislation
and could thus be used by anyone.

Publisher:
Südwestdeutscher Verlag für Hochschulschriften Aktiengesellschaft & Co. KG
Dudweiler Landstr. 99, 66123 Saarbrücken, Germany
Phone +49 681 37 20 271-1, Fax +49 681 37 20 271-0, Email: info@svh-verlag.de

Copyright © 2009 by the author and Südwestdeutscher Verlag für Hochschulschriften
Aktiengesellschaft & Co. KG and licensors
All rights reserved. Saarbrücken 2009

Printed in the U.S.A.
Printed in the U.K. by (see last page)
ISBN: 978-3-8381-0902-2

We believe we're gliding down the highway when in fact we're slip sliding away.

P. Simon and A. Garfunkel

Contents

1	**Introduction**	**3**
	1.1 Outline	4
2	**General Aspects of Terrestrial Atmospheres**	**5**
	2.1 Structure of the atmosphere	5
	2.2 Neutral Middle Atmosphere	7
	2.3 Ionospheres	9
	2.4 The Ozone layer	13
	2.5 Dynamics of the middle atmosphere	15
3	**Coupling of the Sun-Earth System**	**19**
	3.1 The Sun	19
	3.2 Solar particle events	21
	3.2.1 Solar Proton Events (SPE)	22
	3.2.2 Energetic electron precipitation (EEP)	24
	3.3 Solar phenomena	27
	3.3.1 Solar wind	27
	3.3.2 Coronal Mass Ejection (CME)	28
	3.3.3 Solar flares	30
	3.3.4 Corotating interaction region (CIR)	31
	3.4 The Earth's Magnetic field (EMF)	32
	3.4.1 Particle motion	35
4	**Instruments**	**41**
	4.1 Halogen Occultation Experiment (HALOE) on UARS	41
	4.1.1 HALOE Error Estimates	44
	4.2 Geostationary Environmental Satellites (GOES)	49
	4.3 Polar Operational Environmental Satellite (POES)	50
	4.4 Ground-based measurements: Ap-index	52
5	**Solar Proton Events (SPE)**	**53**
	5.1 Properties of SPEs	53
	5.2 Effects on the middle atmosphere	56
	5.2.1 Nitrogen compounds	56
	5.3 Results (SPE)	61

	5.3.1	Long term observation of NO_x	61
	5.3.2	Long time observation of HCl	68
	5.3.3	Bastille Event - NO_x	69
	5.3.4	Bastille Event - O_3	73
	5.3.5	Bastille Event - HCl	74
	5.3.6	Temperature Variation	77
	5.3.7	SPE - Summary	83

6 Model vs. Observation — 85
6.1 The Hybrid Ion Model (HIM) 85
6.1.1 Results for NO and O_3 86
6.1.2 University Bremen Ion Chemistry Model (UBIC) 89
6.1.3 Results HCl 91
6.1.4 Model Errors 92

7 Highly Energetic Electron Precipitation (EEP) — 95
7.1 Electron fluxes: GOES, POES and Ap-index 95
7.2 EEP direct effect 98
7.2.1 EEP long time observation 104
7.3 Indirect Effects (IE) 107
7.3.1 Indirect Effects (IE) and NO_x correlation 110
7.3.2 Indirect Effects (IE) and O_3 correlation 110
7.4 Discussion EEP 114

8 Summary and Conclusion — 117

A Appendix — 119

B References — 124

Abstract

The chemical composition of the middle atmosphere can be strongly influenced by Solar Proton Events (SPEs) and Energetic Electron Precipitation Events (EEPs). These events are well known sources of NO_x (N, NO, NO_2) and HO_x (H, OH, HO_2), which both contribute to ozone loss in the middle atmosphere. Due to its long lifetime, significant amounts of NO_x produced by large particle events in the mesosphere and the upper stratosphere can be transported down into the middle and lower stratosphere during polar winter, where NO_x is a key species in ozone loss. Thus large particle events can potentially contribute significantly to stratospheric ozone loss. This study uses measurements of the Halogen Occultation Experiment (HALOE) instrument onboard the UARS satellite covering the years 1991 - 2005, to investigate mesospheric NO_x production during more than one solar cycle. Furthermore the effect on other species e.g. HCl, which is an inactive reservoir for Cl, and its contribution to stratospheric ozone loss is investigated as well. A decrease of HCl during the SPE in July 2000 could be observed for the first time. Comparisons with the UBIC model, developed at the University of Bremen, showed good correlation with the HALOE HCl data set. Furthermore, an increase of temperature in the thermosphere and a possible decrease at altitudes of the upper mesosphere have also been observed during the SPE in July 2000.

Investigation of the EEP direct effect showed that highly energetic electrons do not affect the NO_x production below 80 km. The EEP indirect effect (IE), in contrast, was found to play a major role in terms of stratospheric ozone loss causing strong inter annual changes during polar winter in the southern as well as in the northern hemisphere.

Further, data of highly energetic electron flux measurements of the POES and GOES spacecraft were compared to the Ap-index to find the best proxy for investigations of particle events affecting the polar regions.

1. Introduction

The coupling between the sun and our planet is one of the most important topics of our century. Not only because of current manmade problems like air pollution but also because of many natural processes affecting the Earth environment ever since. Observation of the sun and the earth atmosphere in the last centuries increased our knowledge and led to some understanding of natural coupling processes like transport, dynamics and chemical properties of the atmosphere. To improve the knowledge of these important topics, this study focuses on impacts of the solar variability regarding the chemical composition of the middle atmosphere in the framework of the international project: Climate And Weather of the Sun-Earth System (CAWSES).

It is well known that the middle atmosphere is influenced strongly by solar variability, especially in the UV range. These changes are transferred directly into variabilities of the photolysis rates which influence the atmospheric composition in terms of the partitioning between short lived species as well as the life time of long lived tracers. Thus, ozone is directly affected mainly due to photolysis in the Hartley bands ($\approx$ 300 nm) as well as the photolysis of O_2 in the Schumann - Runge and Herzberg- continuum bands ($\approx$ 200 nm) which is the main source of odd oxygen in the middle atmosphere. Beside the influence of the UV radiation changes, precipitation of energetic particles of extra terrestrial origin and trapped ones in the magnetosphere also affect the middle atmosphere in many ways. It is well known that cosmic rays, originating from outside our solar system can reach energies up to several GeV and are also able to affect the atmosphere down to the troposphere. Energetic particles, created during massive solar eruptions e.g. solar flares, coronal mass ejections, are accelerated up to several MeV - GeV and can penetrate deep down to the stratosphere in the polar cusp regions, where the field lines are open and connected with the interplanetary magnetic field (IMF). Among these highly energetic particle events two kinds of energetic events have come to scientific interest, as a possible correlation with ozone depletion was observed. These phenomena are called Solar Proton Events (SPE) and highly energetic electron precipitation events (EEP).

SPEs, which are characterized by a sudden increase of highly energetic protons flux, have become a well known source of reactive $NO_x (= N, NO, NO_2)$ and $HO_x (= H, OH, HO_2)$ and were found to be a major source of ozone destruction due to catalytic cycles already by the pioneering work of Swider and Keneshea [1973], Rusch et al. [1981] and Crutzen et al. [1975]. The production of NO_x due to large SPEs and its correlated ozone depletion has been observed several times during the last three solar maxima and can be reproduced by chemical models reasonably well e.g by Jackman et al. [2001]. Further it was found, that compared to quite short lived HO_x which recovers very fast in the polar middle atmosphere, NO_x is long lived. Thus, it can be transported downward into the lower stratosphere during polar winter time, where it is effective for ozone loss in spring and summer time [Jackman et al., 2000].

EEPs occur, in contrast to SPEs, very often during solar minimum, and are characterized by an sudden increase of highly energetic electrons with energies up to several MeV. These events seem to happen due to the acceleration of electrons trapped in the Earth magnetosphere by geomagnetic storms. During strong disturbances, electrons are accelerated to high energies and can precipitate at the free edges of the polar cusps, where the magnetic field lines are open, deep down to the mesosphere. Here they are leading to chemical reactions similar to those caused by SPEs, in terms of NO_x and HO_x production. Studies of these effects have been carried out by Callis et al. [1991], Randall et al. [2007] and Rodger et al. [2008].

However, to obtain more detailed information about the precise coupling processes during and after these charged particle events, more studies with long term data series are carried out in this work.

1.1 Outline

This work presents an investigation of a long time series of data ranging from 1991 to 2005 and focuses on several species e.g. NO_x, O_3 and HCl, located in the middle atmosphere (30 - 100 km), and its chemical response to highly energetic particle events e.g. Solar proton events (SPE) and highly energetic electron precipitation (EEP).

In chapter 1 and 2, the general aspects of the terrestrial atmosphere and the coupling of the sun Earth system are discussed, as well as common solar phenomena and the current knowledge about their origin and dynamics.

Chapter 3 gives an overview of the instruments and data sets used in this study and gives insight into the measurement technique and the reliability of the data.

Results of the SPE investigations and their influence on the chemical composition of the middle atmosphere are summarized in chapter 4 as well as in chapter 5. Here also comparisons with atmospheric models developed at the University Bremen are shown.

Chapter 6 shows results of the EEP direct effect and the EEP indirect effects on the middle atmosphere.

Finally chapter 7 summarizes the main results of this work.

2. General Aspects of Terrestrial Atmospheres

Life on our planet is enabled by our atmosphere and its properties which protects us from solar radiation, has the ability to gain energy by absorption from the solar light, engages stable cycles which lead to an equilibrium in water cycles and works with the electrical and magnetic forces to provide a moderate climate. Due to improvements in our technology the knowledge of our atmosphere has improved greatly in the last 100 years, revealing many secrets of this complex system. This chapter gives a brief overview over some basic aspects of the Earth's atmosphere e.g. structure, main chemical reactions, dynamics and the ozone layer.

2.1 Structure of the atmosphere

The atmospheres of the terrestrial planets (Earth, Mercury, Venus, and Mars) can be structured in different regions with specific chemical and physical properties. The terminology is based on the vertical composition and distribution and the pressure dependency can be described by (Eq. 2.1), derived from the ideal gas law with the quantities pressure p, temperature T, and density ρ and describe the exponential relation between pressure and geometric height **z**

$$p(z) = p_0 \cdot exp\left[-\int_0^z \frac{1}{H(\acute{z})}d\acute{z}\right] \qquad (2.1)$$

$$H = \frac{kT}{mg} \qquad (2.2)$$

Focusing on the thermal properties one can divide the different layers as can be seen in Fig. 2.1. The troposphere forms the lowermost part of the atmosphere. In this region the planetary surface is the primary heat source and the heat is convected by turbulent motion. Hence an adiabatic temperature distribution occurs. The vertical temperature gradient depends on the composition of the atmosphere and on the gravitational acceleration of the planet. The Troposphere ends at the tropopause and is followed by the stratosphere which was thought to be isothermal. But in case of Earth the temperature

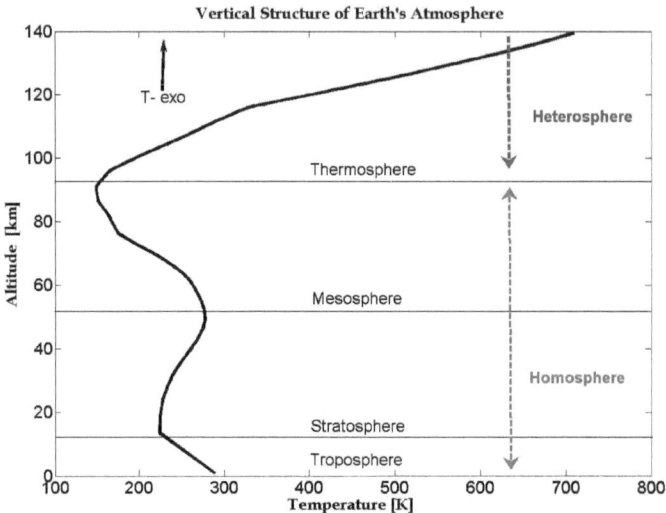

Figure 2.1: *Vertical profile and general structure of terrestrial atmosphere. The temperature profile (solid line) is derived from HALOE data.*

increases due to ultraviolet (UV) absorption processes by Ozone (O_3) and the maximum temperature is reached at the end of this layer, the so called stratopause. Above this, the mesosphere arises where $\partial T/\partial z < 0$ is valid, reaching a temperature minimum at the mesopause. In the Earth's atmosphere the presence of CO_2 and H_2O provides a heat sink by radiating in the infrared in this region. The thermosphere is the next layer and starts after the mesopause terminating the mesosphere. In the thermosphere a positive temperature gradient $\partial T/\partial z > 0$ is reached. Due to solar X-ray and extreme ultraviolet radiation (XUV: 0.1 - 100 nm), heating in this region is engaged. Convection is the major process of heat transport in the lower part of the thermosphere. With rising altitude convection decreases and is finally replaced by conduction, leading to an isothermal region (T=const) the so called thermopause. At the thermopause the mean free path of the gas becomes very large and collisions between the molecules become negligible. Hence light atmospheric constituents can reach velocities to escape the planetary atmosphere and the exosphere starts at the so called exobase or the critical level. The exobase is defined as the level where the mean free path is equal to the local scale height, i.e., the logarithmic decrement of pressure with altitude, shown in Eq. (2.3)

$$\int_{z_c}^{\infty} \frac{dz}{\lambda(z)} = \int_{z_c}^{\infty} n(z)\sigma dz = n_c H \sigma \equiv \frac{H}{\lambda} = 1, \text{ and} \tag{2.3}$$

$$\lambda = (n\sigma)^{-1}, n_c = (\sigma H)^{-1} \tag{2.4}$$

Here λ is the mean free path, n the total number density, σ the gas-kinetic collision cross-section and H is the so called Scale Height. The exobase may also be called the baropause, since the entire atmosphere below that level is also referred to as the barosphere, i.e., the region where the barometric law holds perfectly. In the exosphere the velocity distribution is non-Maxwellian due to the escape of the high velocity particles and the density does not strictly follow the barometric formula, but has to be derived by considering the individual ballistic components of the atmospheric gas, e.g Chamberlain [1963]. One can also find another way of subdividing the atmosphere. It can be structured into upper, middle and lower atmosphere, the lower atmosphere consisting of Troposphere and the lower Stratosphere. The middle atmosphere is commonly defined as the upper Stratosphere and the Mesosphere and the upper atmosphere is consisting of the regions above the Mesopause. It should be noted that without a stratospheric heat source, such as the absorption of O_3 in the case of Earth, a planetary atmosphere may not possess a stratopause and the region above the tropopause may be called either the stratosphere or the mesosphere. Meteorological phenomena occur in the troposphere in contrast to the upper atmosphere where the Aeronomy takes place. The atmosphere can also be divided in terms of its composition into the homosphere and the heterosphere. These terms are somewhat less frequently used than the nomenclature based on the temperature distribution. The composition of the homosphere is uniform and is characterized by turbulent mixing. In the heterosphere, the composition varies due to dissociation of molecular constituents and also as the result of diffusive separation. If the diffusion decreases, mixing becomes the controlling process and the so called homopause (formerly also called the turbopause) is reached. More precisely, the Homopause can be defined as the level where the Eddy diffusion (mixing) coefficient is equal to the molecular diffusion coefficient. Since these coefficients are different for different constituents, they will have different homopause levels e.g. [Bauer and Lammer, 2004]. In the case of the terrestrial atmosphere the homopause levels occur at altitudes of about 100 km. The homopause concentrations of atmospheric constituents represent important boundary conditions for their distribution in the upper atmosphere and the ionosphere [Kazeminejad, 2005].

2.2 Neutral Middle Atmosphere

The main constituents of the Earth's atmosphere are the well known species molecular nitrogen N_2 ($\approx$ 78.08%), oxygen O_2 ($\approx$ 20.95%) and argon Ar ($\approx$ 0.93%) with respect to dry air up to an altitude of about 105 km, these constituents building 99% of the total atmospheric mass. For the distribution versus altitude of the main atmospheric compounds see also Fig. 2.2. All other species are so called trace gases, the most important in particular are water vapor H_2O, carbon dioxide CO_2, methane CH_4, nitrous oxide N_2O and ozone O_3 [Wallace and Hobbs, 2006]. Focusing on the middle neutral atmosphere (mesosphere and stratosphere) the most important reactions taking place are shown in Eq. 2.5 - Eq. 2.10.

- **Photodissociation**

$$H_2O + h\nu \rightarrow H + OH$$

$$O_2 + h\nu \rightarrow O + O \tag{2.5}$$

- **Constituent exchange**

$$O_2^+ + N_2 \rightarrow NO + NO^+ \tag{2.6}$$

- **Dissociation/Oxidation**

$$H_2O + O \rightarrow 2OH \tag{2.7}$$
$$N_2O + O(^1D) \rightarrow 2NO \tag{2.8}$$
$$CH_4 + 2O_2 \rightarrow CO_2 + 2H_2O \tag{2.9}$$

- **Three body reactions**

$$O + O + M \rightarrow O_2 + M$$

$$O + O_2 + M \rightarrow O_3 + M \tag{2.10}$$

In the regions of the mesosphere and stratosphere, the families of odd nitrogen, odd oxygen and odd hydrogen e.g. (N, NO, NO$_2$), (O, O$_3$), (H, OH, HO$_2$) are very important species, as they are involved in catalytic cycles responsible for the balance of the ozone layer. As can be seen in Eq. 2.5 - Eq. 2.10, nitrogen is mainly the result of reactions with molecular oxygen whereby odd hydrogen species are mainly produced by consumption of water vapor. In case of trace gases the standard notation of the so called volume mixing ratio v_i (also referred as mole fraction) turned out to be very useful because adiabatic transport processes preserve the volume mixing ratios [Andrews, 2000]. To derive the volume mixing ratio we consider a small sample of air with volume V, temperature T and pressure p, containing a mixture of gases G$_i$ ($i = 1, 2, 3, ...$). The total number of molecules in this parcel follows by $n = \sum n_i$. Taking into account the ideal gas law in the form

$$pV = nkT \qquad (2.11)$$

where k is the Boltzmann's constant, the partial pressure p_i, which is the pressure of the gas G_i that would be exerted by its molecules under the temperature T and volume V can be derived. Similar considerations apply to the partial volume V_i

$$p_i = n_i \frac{kT}{V} \qquad (2.12)$$

$$V_i = n_i \frac{kT}{p} \qquad (2.13)$$

By knowing this relation we can define the volume mixing ratio v_i as

$$v_i = \frac{V_i}{V} \qquad (2.14)$$

The volume mixing ratio (vmr) and the mass mixing ratio (especially for chemical fluids) are more convenient measures of abundances in case of investigations of transport processes of chemical species, as they are not affected by volume changes. Due to this property the mixing ratio notation enables a clear view especially focusing on chemical production and loss rates. The units are given by ppm (parts per million) ppb (parts per billion) and ppt (parts per trillion) [Andrews, 2000]. In case of dealing with stratospheric ozone a very useful general measure has been defined as the column ozone or total ozone. Here the total number N_3 of ozone molecules in a vertical column of atmosphere of unit horizontal cross section is integrated:

$$N_3 = \int_0^\infty [O_3] dz, \ [cm^{-2}] \qquad (2.15)$$

Further a convenient measure is defined as the height of the column in hundredths of a millimeter. It is assumed that all ozone molecules in the column are brought to a pressure of 1 atm and a temperature of 0 °C. This measure is called Dobson Unit DU. Typical values of column ozone are $\approx$ 300 DU, which means that if the height of the column is compressed, it would be about 3 mm. As most of the ozone molecules are placed in the stratosphere, this measures give good information of the state of the ozone layer.

2.3 Ionospheres

The ionosphere is a very important region of the upper atmosphere. This region is characterized by the presence of charged particles (electrons and ions) of thermal energy. These components are the result of interaction of the neutral atmospheric constituents with electromagnetic and corpuscular radiation. The lower boundary of the ionosphere, which is by no means sharp, coincides with the region where the most penetrating radiation (generally cosmic rays) produce free electrons and ions. The upper boundary of the ionosphere is directly or indirectly the result of the interaction of the solar wind flow outward from the sun at supersonic speeds with the planet's outermost layer. In the case of the Earth we have an intrinsic magnetic field protecting us from direct interaction of the solar wind and the atmosphere, but for weakly or essentially non-magnetic planetary bodies (Venus, Mars, Saturn's satellite Titan), the interaction region between the solar wind and the ionospheric plasma represents the termination of the ionosphere on the sunward side; it is called the ionopause. On the night side the ionosphere can extend to greater distances in a tail-like formation, representing the solar wind shadow. In the tail the extent of the ionosphere is limited by the condition for ion escape. The ionospheres of Venus and Mars are basically different from the Earth ionosphere. Especially the shape of the Earth's ionosphere is dominated by the presence of more complex shape and is structured in different layers, E, F, F1, and F2 with different physical properties and maxima in electron densities.

Earth's atmosphere is a mixture of gases, mostly nitrogen and oxygen. At the surface, nearly all of these gases are present in molecular form (i.e. O_2, N_2). As the altitude above the Earth increases, the density of the gases decreases rapidly and the make up of the gases also changes as some of the molecules are ionized and lead to abundances of e.g. N_2, NO, O_2, N, O and H ions. In contrast, in the atmospheres of Mars and Venus the entire chemistry is dominated by the presence of CO_2, which implies only one electron density peak in the ionospheric structure at the much lower altitude of about 140 km. In case of the Earth's ionosphere the most dominant ionization process above 60 km is the photo-ionization in the UV and X-ray wavelength range [Wayne, 2000], thus a diurnal cycle as well as varying solar activity is showing great impact regarding the ionosphere's density. In this context, the 11-year and the 27 day solar cycle are very significant and show great influence in contrast to energetic particle impacts which only lead to an increase of ionization at high latitudes mainly at geomagnetic latitudes of $>$ 60°. Here enhanced energetic particle fluxes can enter the atmosphere due the opened magnetic field lines (see chapter. 2.5).

Ionic chemistry differs from neutral chemistry as the motion of ions is dominated by Coulomb forces acting over great distances, therefore ion - ion and ion/neutral reactions are much faster than neutral - neutral reactions. Additionally, the motion of ionic species at high altitudes is determined by the electric and magnetic field structure in contrast to neutral species, which are mainly driven by pressure and temperature motion [Wayne, 2000, Scharringhausen, 2008]. Important ionic molecules in the ionosphere are those of oxygen (O_2), nitrogen (N_2), nitric oxide (NO). But also hydrogen existing mostly in its atomic form engages important reactions in this region, shown in Eq. 2.16 - Eq. 2.23.

- **Charge transfer**

$$N_2^+ + O_2 \rightarrow N_2 + O_2^+$$

$$H^+ + O \rightarrow H + O^+ \tag{2.16}$$

2.3. Ionospheres

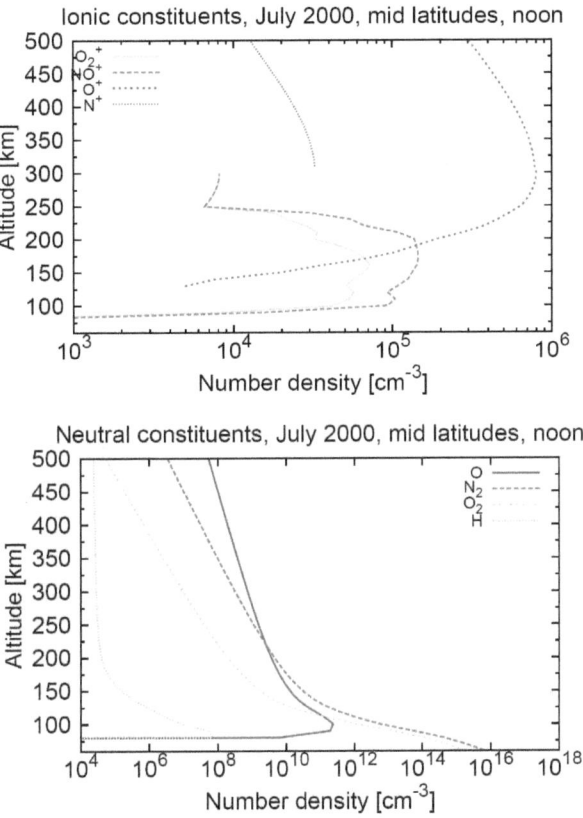

Figure 2.2: *Vertical profiles of major neutral and ionic species in the mesosphere and lower thermosphere. The abundance of neutral species is decreasing with altitude, whereas the ions exhibit maxima at altitudes between 100 and 500 km. As can be seen from the lower panel, the region between 250 and 300 km can be interpreted as the transition between the molecular and the atomic regime. Data acquired from NASA Goddard Space Flight Center. Plots courtesy of Dr. Marco Scharringhausen adapted from Scharringhausen [2008].*

- **Constituent exchange**

$$N_2 + O^+ \rightarrow N^* + NO^+$$

$$O_2^+ + N_2 \rightarrow NO + NO^+ \quad (2.17)$$

- **Photo & Dissociative recombination**

$$O^+ + e^- \rightarrow O + h\nu$$

$$NO^+ + e^- \rightarrow N^* + O \qquad (2.18)$$

- **Attachments** (source of negative ions)

$$O + e^- \rightarrow O^- + h\nu$$

$$O_3 + e^- \rightarrow O^- + O_2 \qquad (2.19)$$

- **Detachments** (sink of negative ions)

$$O_2^- + O \rightarrow O_3 + e^-$$

$$O^- + e^- \rightarrow O + e^- \qquad (2.20)$$

- **Photo dissociation**

$$O_3^- + h\nu \rightarrow O^- + O_2 \qquad (2.21)$$

- **Photo detachment**

$$O_3^- + h\nu \rightarrow O_3 + e^- \qquad (2.22)$$

- **Clustering**

$$O_2^+ + O_2 + M \rightarrow O_2^+ O_2 + M$$

$$H^+(H_2O)_n + H_2O + M \rightarrow H^+(H_2O)_{n+1} + M \qquad (2.23)$$

Due to these chemical reactions the ionosphere is structured into different regions with different properties (see Fig. 2.2). At altitudes of the F-region (above 200 km) therefore mainly O^+ ions are formed and molecular species are rare because of dissociation by short wavelength radiation originated at the sun. Within the thermosphere, where the E-region is located (100 - 200 km), we find in contrast to the F-region ions in molecular form. The dominating species in this region are O_2^+ and N_2^+. The ion chemistry at altitudes below 100 km and thus the mesosphere and upper stratosphere, is very complex as it is a transition region between neutral atmosphere and ionosphere (E and D region) forming large positive and negative cluster ions like $H^+(H_2O)_n$, $NO_3(HNO_3)_n$, etc. Neutral dynamics and influence from electric and magnetic field sources are forcing this region. Hence, clustering and the lifetime of several species due to strongly absorbed radiation have to be taken into account.

2.4 The Ozone layer

A special feature of the Earth's atmosphere is of course the stratosphere and its embedded "ozone layer". The bulk of atmospheric O_3 is located between 20 - 45 km in terms of volume mixing ratio (see Fig. 2.3) and at altitudes of about 15 - 35 km in terms of concentration (see also section. 2.2), in the so called "ozone layer". Typical values of the ozone peak are 4 to 11 ppm and in terms of the column ozone between 250 - 450 DU, depending on season and latitude. Fig. 2.3 shows the different ozone vmr profiles of different measurements carried out by HALOE from February to April 1998 at latitudes from 60° south to 60° north. Clearly the different ozone vmr can be seen with an averaged maximum peak at about 32km and $\approx$ 8 ppm (thick sold line). But also an secondary peak in vmr at altitudes between 80 - 90 km can be seen which may be explained by including odd hydrogen catalytic cycles [Degenstein et al., 2005]. The importance of the stratospheric ozone layer can be roughly summarized in three points:

- A protective shield is formed that reduces the intensity of UV radiation ($\lambda = 0.23$ - 0.32 μm) emitted from the sun reaching the Earth's surface, which enables life on Earth.

- Due to the absorption of UV radiation by ozone an increase of the temperature occurs and can be observed in the vertical temperature profile e.g. in the middle and upper atmosphere. Ozone is the most variable contributor to diabatic heating and cooling of the stratosphere. Thus, the variability of stratospheric ozone impacts the stability of air motion perpendicular to the isentropes.

- Providing the important species O_3, which is involved in many chemical processes.

The first explanation of the O_3 layer and its chemical properties was given by Chapman [1930]. He pointed out, by using a simple chemical scheme considering only oxygen reactions, how the steady state conditions could be preserved in the stratospheric ozone

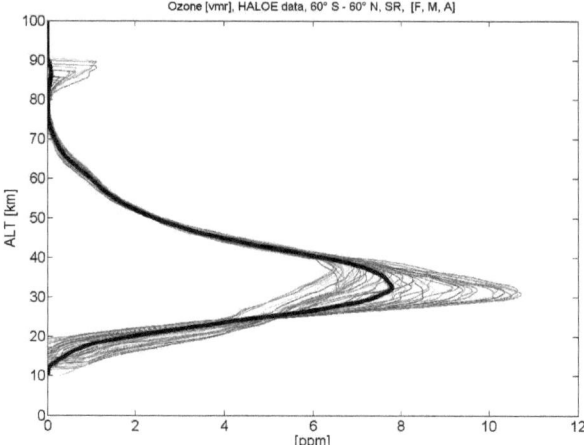

Figure 2.3: *Daily averaged ozone vmr profiles obtained from HALOE data, covering latitudes from -60 ° to 60 ° during the months February, March and April 1998. The black thick line shows the averaged values over all measurements.*

budget. Chapman postulated a set of equations starting with the dissociation of O_2 by solar UV radiation with wavelengths $\lambda < 242$ nm

$$O_2 + h\nu \longrightarrow 2O \qquad (2.24)$$

The released oxygen is mostly in the form of atoms in ground state (^3P) for wavelengths larger than 175.9 nm. At wavelengths below this limit also oxygen in the excited (^1D) is produced. Of course photolysis of ozone occurs too in this cycle, possible for all wavelengths smaller than 1123 nm, with most respect to the interval λ= 230-320 nm where $O(^1D)$ is released. Further atomic oxygen reacts with molecular oxygen to form O_3, where M is an arbitrary molecule which satisfies conservation of momentum and energy, typically N_2 or O_2 molecules are involved.

$$O + O_2 + M \longrightarrow O_3 + M \qquad (2.25)$$

The cycle is closed by the reaction of O_3 and atomic oxygen producing again molecular oxygen

$$O + O_3 \longrightarrow 2O_2 \tag{2.26}$$

As noted before, the Chapman cycle is a simplified model for explaining the existence of the ozone layer. But the Chapman theory predicts too much ozone in the stratosphere. The reason for this overestimation is the neglected presence of ozone destroying catalytic cycles by other atmospheric substances described by:

$$X + O_3 \longrightarrow XO + O_2$$

$$\underline{XO + O \longrightarrow X + O_2} \tag{2.27}$$

$$\textbf{Net: } O + O_3 \longrightarrow 2O_2$$

The net effect is the same as in the Chapman cycle (see Eq. 2.26), here the molecule X is a free radical catalyst, which means it takes part in the reaction but is reproduced in the second step. Very important catalysts to be mentioned in the stratosphere and mesosphere are the species H, OH, NO, Cl and Br. Including these species and their proceedings in catalytic cycles into the theoretical calculation, observation and calculation of ozone distribution are in a good agreement. As several other species are interlinked in odd oxygen destroying reactions and because of their rapid inter-conversions with related species, they are summarized to so called 'families'. In these terms the families of $HO_x = \{H, OH, HO_2\}$, $NO_x = \{N, NO, NO_2\}$, $ClO_x = \{Cl, ClO, 2 \times Cl_2O_2\}$ and $BrO_x = \{Br, BrO\}$ are the forcing species in the stratosphere. Further, potential catalysts can be bound in the form of unreactive 'reservoir species' e.g. dinitrogen pentoxide N_2O_5, chlorine nitrate $ClONO_2$, hydrochloric acid HCl and HNO_3. (Studies of HCl are shown in section 6.1.3.)

Lary [1997] calculated the effectiveness of several catalytic cycles affecting the lower and middle atmosphere. Of great importance in this study are the ozone destroying NO_x compounds, which reach a peak of effectiveness in the middle stratosphere at about 35 km altitude and the HO_x compounds, which reach a peak of effectiveness at about 65 km. Those cycles are very important because the production of e.g NO_x, HO_x and ClO_x increases especially during solar energetic particle impacts, thus leading to an increased source of ozone destroying radicals.

2.5 Dynamics of the middle atmosphere

Beside the chemical cycles also transport processes have to be taken into account. These processes are involved in the formation of the stratospheric ozone layer as well as for the distribution of all species due to air motion by wind fields. One of the most important transport processes in the stratosphere is the Brewer-Dobson circulation (see Fig. 2.4).

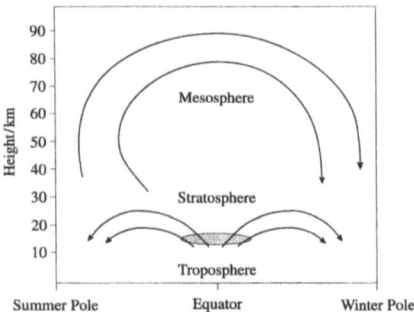

Figure 2.4: *A schematic sketch of the main meridional circulation. The Brewer - Dobson circulation in the stratosphere and solstitial mesospheric circulation. Picture adapted from Andrews [2000].*

This is a meridional circulation in north, south and vertical direction which involves an upward motion of air parcels from the troposphere into the stratosphere at low latitudes, as well as poleward motion in the lower stratosphere of each hemisphere and even a low descent back into the troposphere at middle and high latitudes [Andrews, 2000]. Tropospheric trace gases enter into the middle atmosphere in the tropics, from here they are transported to regions of middle to high latitudes. On the other hand, thermospheric trace gases can enter the middle atmosphere at regions where the middle atmospheric air is transported downward into the troposphere, hence chemical species can be shifted from the equatorial region upwards to the pole regions. At higher latitudes the second major large scale circulation occurs around the solstices. It is characterized by an upward motion in the summer stratosphere ($\approx$ 30-40 km) passing through the mesosphere and descending over the winter pole see Fig.2.4. This process is driven by gravity waves, which are breaking in the mesosphere depositing zonal momentum. This leads to a change in the geostrophic balance including a meridional motion and enables the interaction between different climate regions. Due to the different orography and sea-land coverage of both hemispheres, the meridional circulation deviates, and leads to an stronger planetary wave activity during winter in the northern hemisphere compared to the southern hemisphere. This process, in fact, influences the Brewer-Dobson circulations, taking place below the general circulation at lower altitudes. Hence, both circulations are very important because of their shifting of traces gases and chemical species over long distances. For example O_3 produced at the equator needs typically about only 3 to 4 months to arrive at polar latitudes. Both of these circulations are strongly influenced by gravity waves and planetary waves.

Another important, but latitudinally delimited, transport process are the so called polar vortices which occur in both hemispheres. Strong zonal winds occur especially in the sub-polar region in the winter hemispheres. As a result there are persistent large-scale cyclonic circulations generally centered in the polar winter region. Such a polar vortex seals the encapsulated air masses against meridional inflow, and also weakens the heat flux into the polar regions, especially during winter. However, the Antarctic polar vortex is much more pronounced than its Arctic counterpart which is disturbed by wave activity. As a result, temperatures in the southern polar vortex get much lower than in the Arctic vortex [Andrews, 2000].

3. Coupling of the Sun-Earth System

The sun is the main energy source in our solar system contributing 99,98 % of the total energy that affects the Earth's climate. Thus it has the most dominant effect on the chemical as well as dynamic processes on our planet. The interaction between sun and Earth is a very complex and variable system leading to many phenomena in our atmosphere. This chapter gives an overview of the most important events leading to these intense coupling processes between the most important energy source in our solar system and the Earth's outermost regions, the Earth's magnetosphere. However, it has to be mentioned that many processes have been discovered, but our current scientific knowledge seems to be far away from having substantial explanations.

3.1 The Sun

Our sun is the most important body in our solar system as it enables life on Earth due to its energy supply through radiation. Sunlight is the primary energy source and supplies our planet with $\approx 1370\ Wm^{-2}$ [Aschwanden, 2006]. The sun's estimated age is 4.59 billion years, after it was created due to a rapid collapse of a hydrogen molecular cloud. After a long time of evolution she can now be classified as a G2V stellar spectral type star.

A star is structured into different layers, starting with the core which is considered to extend from the center about 0.2 R_{Sun} with a total density of about 150,000 kg m^{-3} reaching temperatures close to 13×10^6 K. Due to these extreme conditions nuclear fusion sets in by proton-proton chain reactions, where hydrogen is converted into helium and the Bethe-Weizsaecker-cycle or CNO-cycle, where helium reacts with the heavy nitrogen and carbon cores. The proton-proton chain reaction is the major process in our sun, the Bethe-Weizsaecker-cycle only contributes only $\approx 1.6\%$ to the solar energy budget, as it is more effective in stars with more mass. It is important to mention that the core is the only region where reasonable amounts of energy are produced via fusion, the other regions of the star are heated by energy that is transported to the outward layers.

Between 0.2 - 0.7 R_{Sun} from the center the radiative zone is located (see Fig. 3.1). Here heat is transported by thermal radiation, hydrogen ions start to emit photons, which are reabsorbed again after traveling short distances. Thus, radiation transport turns out to be very slow. The convective zone is the outermost layer. Here the solar plasma is

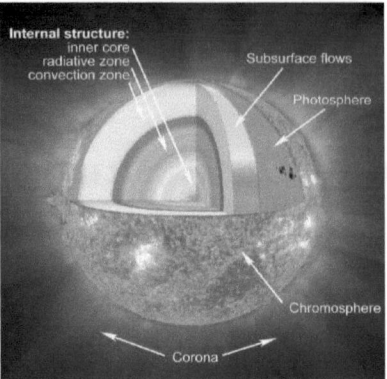

Figure 3.1: *The sun's internal structure, image adopted from www.nasa.gov.*

not hot and dense enough for heat transport to the surface, therefore convection starts, building convection zones and forming the typical granulation imprints on the sun's surface. Due to turbulent convection a small scale dynamo effect is induced, leading to the creation of magnetic fields, similar to the processes driving the Earth's magnetic field, but on a much larger scale. Above this region the photosphere arises, building the visible part of the surface. Here the visible sunlight can propagate into space, emitted due to reaction of electrons with hydrogen atoms. The temperature of the photosphere can be derived by the black body spectrum, leading to a temperature of about 6,000 K [Andrews, 2000]. The sections arising above the photosphere, e.g the chromosphere, the transition region and the corona are commonly mentioned as sun's atmosphere. Those regions reach temperatures orders of magnitudes higher than the sun's surface, increasing gradually with altitude. As these regions are very unstable and changing permanently, they are sources of many solar phenomena e.g solar wind, coronal holes and others.

The sun mainly consists of hydrogen (H) 74,9% and helium (He) 23,8% and trace quantities of iron (Fe), nickel (Ni), oxygen (O), silicon (Si), sulfur (S), magnesium (Mg) etc. Very important for the understanding of many solar phenomena is the very strong magnetic field reaching field strengths in the order of 1000-2000 G [Keller et al., 1990]. As nearly all matter on the sun exists in the form of gas and plasma, the rotation at the equatorial range (about 25 days) is much faster than in polar regions (about 35 days). This differential rotation leads to a very complex shape of the magnetic field, as the field lines become twisted over time, forcing phenomena like changing magnetic activity, sunspots and eruptions at the sun's surface. Further, a self powering of the solar dynamo effect occurs, which is strongly connected to the magnetic field. This property is also known as the 11 year solar cycle, which is actually a 22 year solar magnetic cycle, because after 22 years the magnetic field constellation will be at its starting point again. Its connection to the changing solar activity and altering sunspot number was discovered first by Schwabe [1844]. As the solar magnetic activity changes with a constant period, also changes of the sun's atmospheric structures (e.g corona and wind) are detected. Further a modulation of the solar irradiance, modulation of short-wavelength solar radiation flux from ultraviolet to X-Ray, modulation of the occurrence frequency of flares, coronal mass ejections, and other effective solar eruptive phenomena can be observed (see Fig.3.2).

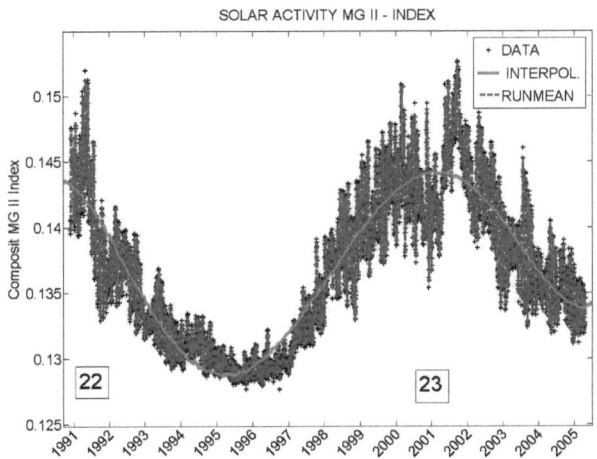

Figure 3.2: *Solar activity from 1991 - 2005, spanning the solar cycles 22 and 23. Solar activity derived from Mg-II index data obtained from Dr. Weber IUP Bremen, data available from the institute homepage: www.iup.uni-bremen.de/gome/. Black dots are daily averaged data, red solid line are interpolated values and blue line shows 3 days running mean values. For further information about the Mg-II index as proxy for UV solar flux variability associated with the 11-yr solar cycle see Weber [1999].*

3.2 Solar particle events

Apart from changes of the solar irradiance, one of the most dramatic impacts of solar variability affecting the Earth and the whole solar system, are solar particle events, which from the point of view of our planet, are large and sporadic increases of energetic particles from the sun. Already the name Solar Energetic Particle Event (SEP) suggests that a very wide range of particles are included in a massive process. Due to various processes induced by a very hot and magnetized corona and the interplanetary plasma, a population of highly accelerated charged particle is created and ejected. The energy ranges of this plasma, mainly consisting of protons, electrons and a variable admixture of He^{2+} and even heavier nuclei like Fe (private conversation with Dr. Ilan Roth, University Berkley), reaches from several keV to some GeV. Particles are sometimes accelerated up to half the speed of light, therefore reaching the Earth magnetic field within hours to minutes. First necessity for investigation of solar particle events arose due to concern about safety for humans and technical systems during space missions outside the protecting magnetosphere. The biggest SEPs since detailed observations started by the National Oceanic and Atmospheric Administration (NOAA) in 1976 were detected on October 1989, March 1991, July 2000, October 2003 and November 2004. A detailed list of a special case of Solar particle events which are dominated by a high proton flux can be found at: http://www.sec.noaa.gov/ftpdir/indices/. The processes leading to the formation of these SPEs are not yet clear, but massive events like CMEs, flares and solar wind processes (mentioned in sections 3.3.1 - 3.3.4) show a good correlation with these events, thus these features seem to be some of the main driving forces of SEPs.

This study focuses on two special particle events. On the one hand the coupling of the middle atmosphere with highly energetic electrons called high energetic electron precipitation (EEP), and on the other hand on the appearance of massively increased high energetic proton fluxes, commonly called Solar Proton Events (SPE). Both kinds of events seem to originate in solar particle events.

3.2.1 Solar Proton Events (SPE)

During the solar maximum the sun's most outer layers are very active, thus leading to an increase of sunspot number, flares, coronal mass ejections (CME) and eruptions at the surface and the corona. Generally we can distinguish the slow solar wind reaching velocities in the range of $\approx$ 300 - 500 kms^{-1} and the fast solar wind with velocities of up to $\approx$ 500 - 800 kms^{-1}. Phenomena like coronal mass ejections can even accelerate particles up to 2500 kms^{-1}. In case of SPE there are three main solar processes leading to an increase of higher proton fluxes (private conversation with Dr. Ilan Roth, University Berkley). At first, impulsive events e.g solar flares, and secondly, gradual events e.g coronal mass ejections (CME) and (CIR) corotating interaction regions. Solar flares are called impulsive events because they last only in the range of hours to minutes. Very rapidly electrons, protons and other nuclei (e.g Fe, ^3He) are ejected and accelerated to energies up to several MeV. CMEs are gradual events and seem to be correlated highly to the 11 year solar cycle (see Fig. 3.2 and Fig. 3.4). Large amounts of charged particles are accelerated up to > 15MeV while surfing on the stream reaching velocities up to 2500 kms^{-1} [Roth and Bale, 2006]. These events can last for several days and in case of the SPE in October 2003 particles only took $\approx$ 19 hours to reach the Earth's magnetosphere [Mewaldt et al., 2005]. Due to its supersonic speed a shock is formed in front of the stream approaching the Earth magnetosphere. As a consequence of the high energies and velocities, protons are able to penetrate deep into the atmosphere within the polar cusp regions where the magnetic field lines are open. Another source for SPEs are the corotating interaction regions (CIR). This phenomenon occurs if slow solar wind is rammed by fast solar wind and is correlated to the 27 days solar period.

3.2. Solar particle events

Also in this case the particles are accelerated to several MeV. Due to these large solar storms an increase of high energy proton flux towards Earth is observed, e.g. by the Geostationary Operational Environmental Satellite (GOES) (see Fig. 3.4). The definite origin of Solar Proton Events (SPE) is still not very well understood. Recent studies show a complex interaction between many solar activity phenomena like Coronal Mass Ejections (CME), solar flares and even radio type II emissions [Wang, 2006]. As the solar wind enriched with highly energetic particles approaches the Earth the magnetosphere is shaped due to the high flow speed and density. These protons are guided by the Earth's magnetic field towards the middle atmosphere [Jackman et al., 2005b].
Further protons with energy levels between 10 - 100 MeV impact the neutral middle atmosphere (mesosphere and stratosphere) and lead to ionization, dissociation, dissociative ionization and excitation. The penetration depth of the highly energetic particles is depending on their energy level, as they gradually lose energy along their trajectory through the atmosphere, see Fig. 3.3.

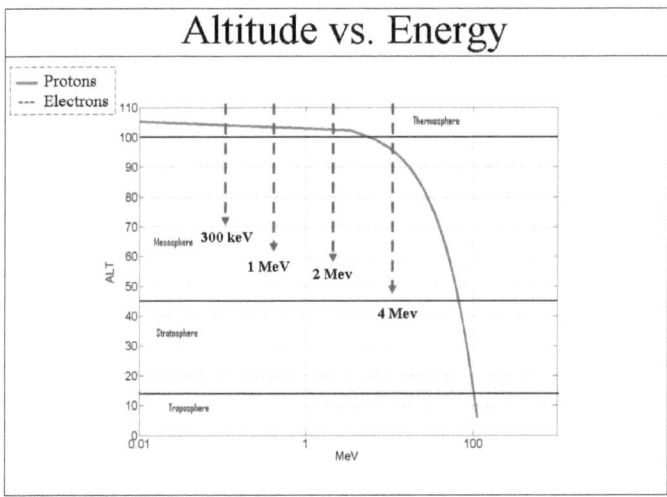

Figure 3.3: *Required energy of protons and electrons for penetrating different atmospheric layers. Altitude in km, energy in eV. Proton are represented by red solid line and electrons by blue dashed arrows. Image derived from data obtained from Vogt et al. [2008].*

In case of investigating the middle atmosphere, protons with energies between 1 - 100 MeV have to be considered. These protons can arrive very fast at the Earth's magnetic field because of their very high velocities. Protons with energies $\geq$ 40 MeV can penetrate very deep into the atmosphere and then deposit their bulk of energy in the polar stratosphere. In contrast to highly energetic protons, slower moving protons with a lower energy level (with energies near 1 MeV) can only penetrate down to the region

Table 3.1: *Peak of ionization rates by protons with several energy levels. Data courtesy of Clilverd [2008].*

Energy [MeV]	Altitude [km]	Ionization rate $[cm^{-3}s^{-1}]$
1	90	1×10^{-1}
2	85	2×10^{-1}
4	80	3×10^{-1}
10	70	9×10^{-1}
20	60	1×10^{0}
40	50	2.5×10^{0}
100	40	6×10^{0}
200	30	1.5×10^{1}
400	20	3×10^{1}
1000	10	7×10^{1}

of the thermosphere and upper mesosphere [Jackman et al., 2005b]. For investigations of the influence of SPEs on the mesosphere, these slower protons are important. These particles with an energy range of about 1-30 Mev deposit their bulk of energy into the mesosphere. A very active solar period occurred between July 2000 and November 2001 at the peak of the solar maximum of the solar cycle 23 and in October 2003 approaching the end of solar maximum. Very strong SPEs were measured in July 2000, November 2001 and October 2003. GOES detected proton particle fluxes during the large SPEs, with values over 500 $particles/cm^2/s/sr/MeV$ at 3 energy level channels from 8 MeV to 100 MeV (see Fig. 3.4).

3.2.2 Energetic electron precipitation (EEP)

A sudden increase of highly energetic electrons near Earth, also called EEP or REP for relativistic electron precipitation event because of their high velocities (see also Appendix A), is a phenomenon that is monitored regularly by several spacecrafts which are observing precipitating particles in polar orbits e.g POES (see Fig. 3.4). EEPs are forced due to magnetospheric appearances which are caused by substorms as a result of the variability and extreme and sporadic events of the sun. Substorms are leading to disturbances throughout the magnetotail and further cause effects that are prejudicing the whole Earth's magnetic system [Vagina and Popov, 2000]. Energetic particles in the inner magnetospheric system are forced basically to three types of motion. First, they have to gyrate around the magnetic field lines, secondly they have to oscillate perpendicular to the magnetic field lines between the mirror points and thirdly, a drift of electrons and protons in different directions at the equatorial region. Trapped particles in the Van Allen belt are separated depending on their charge and energy in different shells with different distances to Earth. Where exactly a particle mirrors is determined by its pitch angle (the angle between particle velocity vector and the magnetic field strength), which forces the particle to mirror and to keep trapped. Reaching pitch angles near to 0 ° or 180 °, the particles will be lost because they penetrate deep into

3.2. Solar particle events

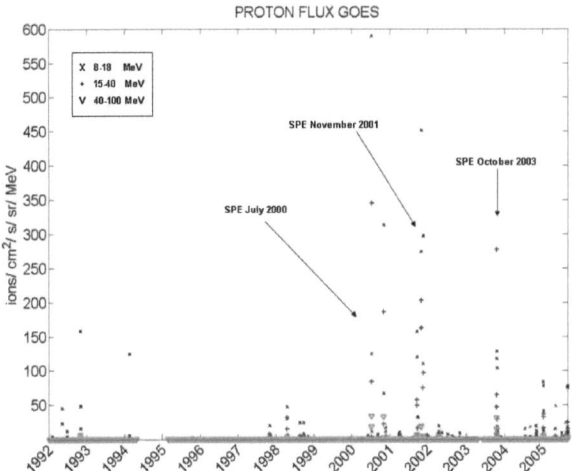

Figure 3.4: *Proton flux measurements provided by GOES for three different channels representing different energy levels. X icons show energy levels from 8 - 18 MeV, + icons shows protons with energy level from 15 - 40 MeV and finally V icons representing protons with energies from 40 - 100 MeV. Plot were derived from data obtained of NGDC, available at: http://spidr.ngdc.noaa.gov/spidr/home.do*

the atmosphere at polar regions [Akasofu and Chapman, 1972]. Exactly these loss processes are happening during disturbed magnetospheric conditions, e.g substorms and whistlers, which are the origin of EEPs. The exact origin of the substorms is still under great discussion but their effects on the magnetosphere can be observed very well. Due to deformation and compression of the magnetic field lines, electrons traveling on outer magnetic shells are forced into inner shells. As the adiabatic invariants have to be conserved, the electrons are accelerated to extremely high energies up to several MeV, therefore they are called relativistic electrons. However, this process is very slow ($\tau \approx days$) compared to the second source of relativistic electrons, whistlers. Whistlers are the dispersed electromagnetic signatures caused by lightning strikes. The pulse of radiation generated by a lightning discharge propagates over great distance in the Earth-ionosphere making its way into the magnetosphere. Here they accelerate the electrons which are trapped to magnetic field lines in the so called van Allen belt, up to several MeV, leading to EEPs. Whistlers cause very rapid reactions, the average time for this process is $\tau \approx minutes$ (correspondence, Dr. Ilan Roth). Whistler induced Electron precipitation (WEP) is a driving process in the inner radiation belt and affects the atmosphere down to altitudes below the ionospheric D-region [Rodger et al., 2005]. Investigations of EEPs and their effects on the middle atmosphere have been presented and discussed in studies like Gaines et al. [1995], Frahm et al. [1997], Callis et al. [1991] or Randall et al. [2007]. In contrast to SPEs that occur mainly during solar maximum (Baker et al. [1986]), EEPs seem to appear mostly during solar minimum conditions, see also Fig. 3.5 derived from GOES and POES data. Here electron fluxes 300 keV $\leq E_e \leq$ 2.5 MeV, proton fluxes 8-14 MeV and 15-44 Mev and the solar cycle derived by the Mg-II index is plotted together. The important SPE's e.g between the years 2000-2002 occur obviously during solar maximum conditions, in contrast to the EEP which appear more frequently during declining and intensifying solar activity. The required energies of electrons to penetrate into the mesosphere (below $\approx$ 100 km) are at least $\approx$ 10 keV, to pass through the stratosphere at least $\approx$ 4 MeV.

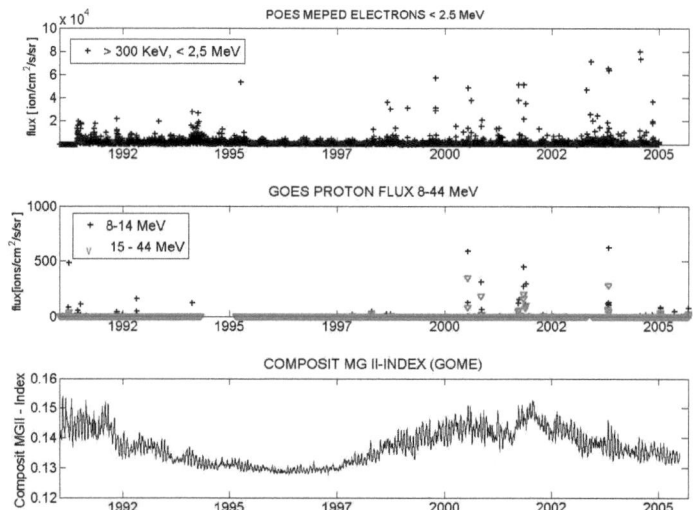

Figure 3.5: *A) POES electron flux measurements covering the years 1991 to 2005. Electron energy levels < 2.5 MeV. Data obtained from Jan Maik Wissing, University Osnabrück, Department for Numerical Physics B) GOES electron flux data covering the years 1991 to 2005. Counts of electron flux with energy levels between 300 keV and 2.5 MeV. C) GOES proton flux measurements for energy levels 8 - 44 MeV for the years 1991 - 2005 D) Solar activity derived from Mg II - index data covering years 1991 to 2005, data obtained from IUP homepage [Weber, 1999].*

Table 3.2: *Peak of ionization rates by electrons with several energy levels [Clilverd, 2008].*

Energy [keV]	Altitude [km]	Ionization rate [$cm^{-3}s^{-1}$]
4	120	5×10^{-3}
10	110	2.5×10^{-2}
20	95	4×10^{-2}
40	90	9×10^{-2}
100	80	2×10^{-1}
200	75	3×10^{-1}
400	70	8×10^{-2}
1000	60	2×10^{0}
2000	55	2.8×10^{0}
4000	50	6×10^{0}
10000	40	2×10^{1}

3.3 Solar phenomena

Solar particle events can lead to an increased flux of highly energetic protons (SPE) as well as to an increased flux of highly energetic electrons (EEP) due to several types of acceleration processes. Both events are influenced by solar activity. This section gives a brief overview of important solar phenomena.

3.3.1 Solar wind

The notion of the solar wind as a continual flow of particles radially emitted by the sun's atmosphere was developed by observations during total eclipses of the sun. Those observations led to the understanding of an open solar atmosphere influencing its surrounding space due to absorbed radiation as well as through ejected matter. Evidence for this hypothesis was gained by the first spacecrafts making measurements outside of the protecting Earth's magnetosphere. Soon observations pointed out the existence of different types of solar wind differing from each other throughout parameters like e.g. densities, flow speed and temperature (see Fig. 3.6). Properties of two different types of solar wind, the low speed wind and the fast wind, are listed in Tab. 3.3.

The solar magnetic field is embedded into the plasma and is pulled radially outward by the expanding solar wind. Because of the sun's differential rotation the field lines and the connected particles are forming the shape of a spiral, also called the Parker-spiral. This model of the heliosphere was very sufficient focusing on the slow solar wind and fast solar winds were assumed to be an exception. In the 1970s coronal holes were discovered when X-ray telescopes were first lifted outside the Earth's atmosphere to reveal the structure of the corona across the solar disc. Coronal holes are associated with open magnetic field lines and mostly located at the sun's poles in contrast to the regions of the equator where solar spots are found but the magnetic field lines are closed. Those regions of the coronal holes are the origin of the fast solar wind [Glassmeier and Scholer, 1991], (see also Tab. 3.3), and lead to the formation of a

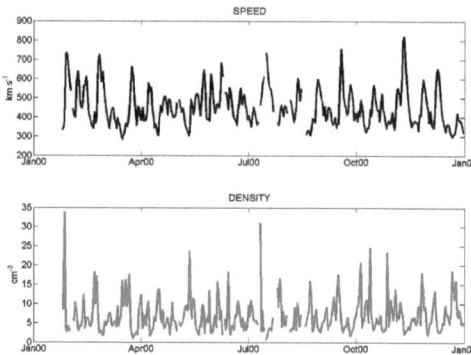

Figure 3.6: *Altering solar wind speed and densities covering the year 2000. Derived from data of the Solar and Heliospheric Observatory (SOHO).*

positively charged north pole and a negative south pole. Due to the opened magnetic field lines originating at polar regions, and the regions around the equator with closed magnetic field lines, a disk-like area arises, separating the positive and negative field lines. Thus, a rotating heliospheric current layer is induced somehow reminding to a dress of a dancer, therefore the name 'Ballerina model', was first proposed by Alfven [1977] (see also Fig. 3.7). Because of the changing activities at the equatorial area e.g sun spots, as well as in the polar regions where the coronal hole activities are altering, the whole magnetic and current system is in permanent alternation. Each time faster solar wind is emitted, fluctuations in the heliospheric system are induced, leading to changes in the geomagnetic field.

3.3.2 Coronal Mass Ejection (CME)

A special feature of the sun are the so called coronal mass ejections (CME). As the name suggests, massive amounts of matter are ejected within a few hours from the corona. Thousands of these massive eruptions have been observed and investigated. Thus averaged data of their properties are well known. The mean velocity of the accelerated matter is about 470 kms^{-1} up to maximum $\approx$ 2500 kms^{-1} (private communication with Dr. Ilan Roth, University Berkley). Matter with masses of 4.1×10^{12} kg are ejected reaching energies of up to $E_{kinetic}$=3.5 $\times 10^{23}$ J. Those impressive phenomena with the shape of a bubble can be observed around the whole solar disk but seem to occur mostly near the equator. A huge catalogue of more than 10.000 CMEs from the past 10 years has been compiled by Yashiro et al. [2004] using data from the Solar and Heliospheric Observatory (SOHO) orbiting at the sun Earth L1 point. A clear description of a typical CME can not be given, some scientists claim that there

3.3. Solar phenomena

Table 3.3: *Averaged solar wind parameters at 1 AU (see Appendix A) during solar activity minimum conditions, compiled by Schwenn [1990].*

	Low speed wind (LSM)	Fast wind (HSS)
Flow speed v_p	250 - 400 kms^{-1}	400 - 800 kms^{-1}
Proton density n_p	10.7 cm^{-3}	3.0 cm^{-3}
Proton flux density $n_p v_p$	$3.7 \times 10^8 cm^{-2} s^{-1}$	$2.0 \times 10^8 cm^{-2} s^{-1}$
Proton temperature T_p	3.4×10^4 K	2.3×10^4 K
Electron temperature T_e	1.3×10^5 K	1×10^5 K
Momentum flux density	2.12×10^8 dyne cm^{-2}	2.26×10^8 dyne cm^{-2}
Total energy flux density	1.55 erg $cm^{-2} s^{-1}$	1.43 erg $cm^{-2} s^{-1}$
Helium content n_{He}/n_p	2.5%, variable	3.6%, stationary

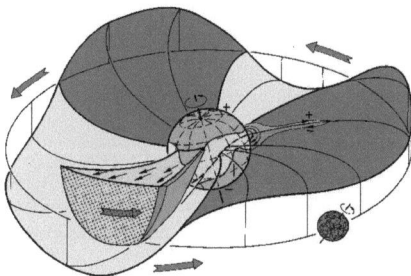

Figure 3.7: *Formation of the heliosphere described by the 'ballerina model' in three dimensions, according to Alfven [1977]. Image adapted from Glassmeier and Scholer [1991].*

are two main types, others claim that there are even more types. However, the two most common types are :

- **Gradual CME:** Plasma emitted with balloon like shape, accelerates slowly over large distances to speeds of 300 - 600 km/s.

- **Impulsive CME:** Often associated with flares, accelerates up to speeds of $\geq$ 2000 km/s.

By now it is not yet clear if those are two fundamentally different processes or if they represent different extremes of the same process [Schwenn, 2006]. The usual initiation of CMEs is described by Zhang et al. [2004] and is separated in a three phase scenario: **Initiation Phase, Impulsive Acceleration phase, Propagation Phase**. To point out a specific origin of the CMEs is not really possible at the current state of knowledge, as mass ejections seem to occur at polar as well as in mid-latitude regions. Also recent studies from Cremades and Bothmer [2004] show that CMEs seem to be deflected from their middle latitude origin towards the solar equator. Also the connection between solar flares and CMEs is not very well understood, as sometimes flares seem to be correlated with masse ejection and sometimes not. Recapitulating we know that CMEs are massive ejections of gigantic clouds of ionized gas (highly diluted plasma) consisting of up to of 30% He^{++}, He^+ and Fe^+_{16}, provoking shock waves and leading to massive interactions in the interplanetary magnetic field.

3.3.3 Solar flares

Solar flares are one of the most dramatic and fascinating eruptions that can be observed and have been known for a long time. The first direct observation was back in 1859 by Carrington [Carrington, 1859]. He discovered the bright flash during his ordinary sunspot observations and published in the same year the connection of this event with the geomagnetic storm which arrived only about 17 hours later. Since then, the coupling of solar activity and geomagnetic field disturbances has become a serious scientific topic also referred to as 'Space weather'. The flare back in 1859 must have been one of the most energetic events ever since, as the ensuing geomagnetic storm induced aurora lights at latitudes as low as 20°, and ground induced currents in telegraph wires in the united states as well as in Europe [Tsurutani et al., 2005].

A flare can be described in general as an energetic fast process which enables to release a package of electromagnetic radiation within seconds to minutes covering a range of wavelengths over several orders of magnitude. Radiation is emitted and can be detected at 1 AU distance from high frequency γ-rays down to low frequency radio bursts with wavelengths with magnitudes of a few km (see Fig. 3.8).

- γ-rays:

This very high frequency ($\approx 10^{20} Hz$) radiation with energies between keV and MeV is produced during very strong flares. There are several physical processes leading to this kind of emissions e.g. electron bremsstrahlung continuum emission, nuclear de-excitation line emission, positron - electron annihilation line emission and others. Some of these processes lead to energies up to several GeV. During flares also protons are accelerated, therefore they can penetrate deep into the solar corona. Here they are forced to excite and γ-rays as well as white light is emitted. To engage this process, proton energies of at least 20 keV are required. Because of the necessity of these high energies, only big flares can be seen in the visible range as an appearance of white light [Schwenn, 2006].

- X-rays:

X-rays are very often the first signature of a solar flare rising up to energies of several keV (soft X-ray, frequency $\approx 10^{16}$) and MeV (hard X-ray, frequency $\approx 10^{19}$). Due to extreme processes in the very hot thermal plasma (T $\approx 10^7$ K) X-ray radiation is released due to the bremsstrahlung continuum and a multitude of lines of heavily stripped ions. In case of the big X-ray flare on 28 October 2003 a total power of X-ray flux of 1.72 mWm^{-2} was measured. So called hard X-ray radiation is released usually a few minutes after the soft X-ray burst. This kind of non thermal radiation arises due to highly accelerated electrons colliding with atoms of lower and thus denser layers of the sun's surface producing a bremsstrahlung continuum in the form of hard X-ray [Schwenn, 2006].

- Extreme Ultraviolet Radiation (EUV):

From the lower level of the solar atmosphere strong emissions in the EUV can be detected. Because of these emissions of EUV radiation lines, we can observe the so called transition layer of the flare, which is also heated up quite abruptly. Especially the Lyman-α line (121.6 nm) and the other members of the Lyman series of hydrogen are emitted very strongly from the chromosphere. In the lower atmospheric layers the Balmer series of hydrogen becomes dominant. The most important is the H-alpha line, which is located near the visible in the solar spectrum. Thus the intensity of the H-alpha line was used for classification of flares, see also Reid [1963].

- Microwave bursts

Due to the coupling processes between fast electrons and the strong chromospheric magnetic fields also microwave radiation is produced. The spectrum is a broadband continuum with peak intensities at some tens of GHz [Schwenn, 2006]. A very good time correlation between hard X-ray and microwave bursts can be observed, especially beyond 1 GHz.

- Radio bursts

Radiation with wavelengths of radio signals are emitted during a flare e.g. type 4, 3, 2, kilometric type 2 (see also Appendix A). The production of this radiation can be explained by a two step process. At first electrons are accelerated by the flare to energies in the range of some keV, exciting plasma oscillations locally on their way from the chromosphere into the heliosphere. Due to the non linear wave-wave interaction the plasma oscillations are converted into electromagnetic radiation, which can be measured by radio antennas. More detailed information about all types of radio bursts can be found in Schwenn [2006].

3.3.4 Corotating interaction region (CIR)

An interesting feature, but not very commonly mentioned, are the so called corotating interaction regions CIR, leading as well to strong disturbances of the geomagnetic field. There are mainly three kinds of interplanetary structure:

- CIR
- CIR/ICME

- ICME

The first type CIR is the 'pure' corotating interaction region type, where a fast stream originated from a coronal hole rams a slow stream in front, associated with the closed streamer belt region. Those phenomena appear when no flares or CMEs are ejected, but are of course strongly related to the 27 rotational day period of the sun. As the flow pattern from the sun is roughly time stationary, the fast and the slow solar wind are forming spirals in the equatorial plane corotating with the sun. The second type is the CIR/ICME (Interplanetary Coronal Mass ejection), where a shock wave is accelerated by a coronal mass ejection and hits the slow solar wind in front. Just like CMEs and flares, this type of CIR is transient. The third type are the 'pure' interplanetary mass ejections, when a gradual or impulsive CME affects the undisturbed interplanetary environment, which is also mentioned in section. 3.3.2. However, CIR are important processes especially during solar minimum, leading to strong activity in the geomagnetic field. Richardson et al. [2000] showed that about 70% of geomagnetic disturbances are caused by CIR, 20% due to strong solar wind and only 10% by CMEs. Recent studies by Zhang et al. [2008] showed big influences of CIR on the Kp index during the solar cycle 23, thus giving good evidence of the importance of this phenomenon.

3.4 The Earth's Magnetic field (EMF)

From the sun a permanent flux of particles (neutral and charged ones) is radially emitted and can be described as highly diluted plasma reaching velocities of over $\approx 1200\ kms^{-1}$ [Raith, 1997]. Because of the strong intrinsic and quasi - dipolar Earth magnetic field which is produced by the so called geodynamo effect, the wind streams can not affect the Earth's atmosphere directly, as the magnetic field (EMF) is acting like a protection shield. Typical intensity of the geomagnetic field at the surface are about 50.000 nT at polar regions and 30.000 nT in the equatorial plane (see also Appendix A). Due to the existence of the EMF the supersonic solar wind stream is slowed down to sonic speed at the so called bow shock in front of the magnetopause (see Fig. 3.9), it is typically formed at $\approx 13.5\ r_E$ depending on the solar wind conditions. Here, the particles are slowed down and have to course around the magnetopause. The solar wind and the magnetospheric plasma are in pressure balance, which arises under common solar wind conditions at about 10 Earth radii. During fast solar wind conditions the magnetopause can be compressed to about 6.6 Earth radii towards earth, because of the high wind pressure [Pulkkinen, 2007]. Because of the very high kinetic pressure on the EMF's day side the field lines are compressed and on the night side the typical shape of the magnetotail is formed. The magnetotail is formed of open magnetic field lines originated at the polar regions and can extent to $3000 r_E$. As coupling processes with the solar wind induce fluctuating magnetic fields, electric currents are also induced. This mechanism follows from the Maxwell equations e.g. $\nabla \times \vec{B} = \mu_0 \vec{j} + \frac{1}{c^2}\frac{\partial E}{\partial t}$ where B describes the magnetic induction, μ_0 is the magnetic constant and j the total current density. Thus, a set of different current systems arises with amperages of millions of Amperes [Glassmeier and Scholer, 1991].

3.4. The Earth's Magnetic field (EMF)   31

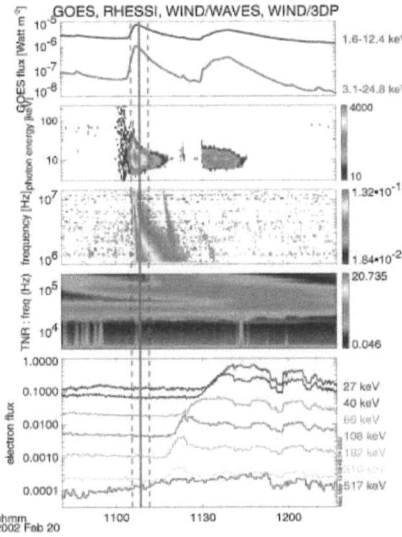

Figure 3.8: *Example of a flare hard X-ray burst observed by RHESSI with corresponding solar type III radio burst and energetic electrons (and Langmuir waves) observed in situ by the WIND spacecraft [Krucker and Lin, 2002]. Top panel: GOES soft X-rays; second panel: Spectrogram of RHESSI X-rays from 3 to 250 keV; third and fourth panels: radio emission observed by the WIND WAVES instrument; fifth panel: Electrons from 20 to 400 keV observed by WIND 3-DP instrument. From Lin [2005].*

The region between the bow shock and the magnetopause is the so called magnetosheath, which is a region full of turbulence because of the incessantly changing direction of the magnetic field and the disorganized plasma stream. The solar wind cannot penetrate the magnetopause because when the plasma is approaching the charged particles are deflected, protons in west direction and electrons eastwards. Hence, a current system perpendicular to the geomagnetic field, the so called Chapman - Ferraro system, forms. This is responsible, with other current systems at the magnetotail, for the shielding of the magnetosphere at its outer layer. The magnetosphere consists of the north lobe and the south lobe with opposite directed magnetic field lines clockwise and counterclockwise. In the middle both currents sum up to the neutral sheet current. At the inner region, depending on the space weather conditions, the so called ring current forms between 2 and 9 r_E at equatorial latitude. The ring current is directed westwards and dominated by protons. Those charged particles with energies up to several keV are trapped in the EMF and are permanently forced to mirror between both hemispheres.

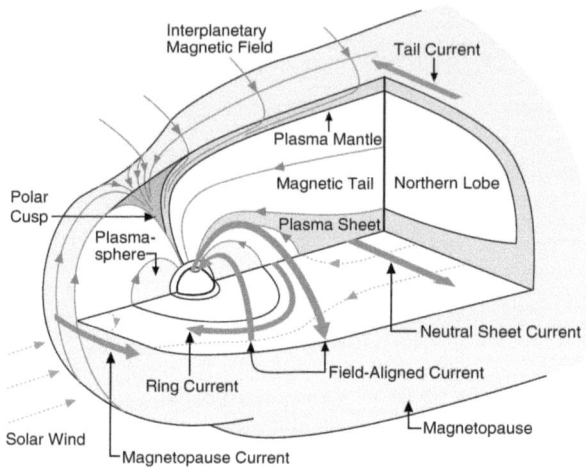

Figure 3.9: *Schematic plot of the Earth's magnetosphere during interaction with the solar wind. Sketch adopted from Russell [2000].*

This mirror effect at the so called mirror points are a consequence of the conservation of energy and the conservation of the magnetic momentum of charged particles. Beside these currents, directed only perpendicular to the magnetic field lines, systems with parallel flowing currents e.g. Birkeland currents arise, connecting the magnetospheric and the ionospheric systems, establishing an exchange of energies between those regions. The forcing power behind these processes is once again the kinetic energy followed from solar wind and the border layers of the magnetosphere. The Ionosphere acts like a resistor, where hundreds of Gigawatts are first converted in the joule heating and then again converted into kinetic energy. The biggest three dimensional current system with perpendicular as well as parallel directions is the polar electrojet induced by the Hall currents flowing eastwards and the Pedersen currents westward. This system arises due to the increased conductivity at polar regions, induced by large scale magnetospheric plasma convection [Glassmeier and Scholer, 1991].

Plasma connected to the Earth's magnetosphere is dominated by electrons and protons originating mainly from the solar wind and the ionosphere. The distribution of the plasma is separated in sharply defined regions depending on its temperature, specific density and distribution function (see Fig. 3.10). Looking at the magnetosphere from the day side towards Earth, one can see that after passing the magnetopause a region

3.4. The Earth's Magnetic field (EMF)

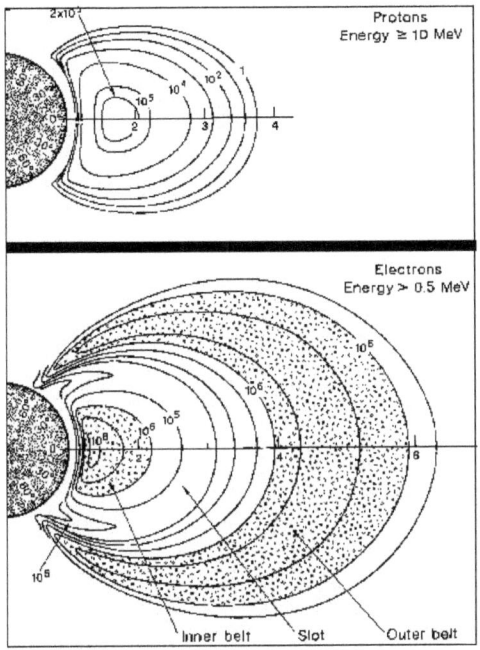

Figure 3.10: *Earth's radiation belts. The numbers give the omnidirectional flux in particles per square centimeter per second. Sketch adopted from Kivelson and Russel [1995].*

with energetic ions and electrons arises with plasma densities of $\approx 10^5 m^{-3}$. Ions in these regions reach energies of several keV and electrons in the range of hundreds eV. Moving further towards Earth, the energies of the particles are increasing up to the ring current. At about 4 Earth radii a sharp border with a shift in particle densities from about 1 particle cm^{-3} to 1000 particle cm^{-3}, the outer layer of the plasmasphere is reached, the so called plasmapause. This region is filled with cold plasma diffusing from the ionosphere along the magnetic field lines and exists only in a torus between 60 ° geomagnetic latitude. Inside the plasmasphere the charged particles are rotating as a fixed body with Earth, outside escaping plasma is immediately transported into higher regions by electric convection fields.

3.4.1 Particle motion

Plasmas are a state of matter, like solids, liquids, and gases, with properties quite distinct from gases. In case of Earth's magnetosphere, most of the plasma originates from the solar wind, as the sun's surface supplies ideal conditions for its production.

The fluid states of gas and liquid are described by the Navier-Stokes equations whereas plasmas are described by the Maxwell, Boltzmann, and Vlasov equations. The environments where plasmas are found are so severe that neutral components (atoms, molecules) found in gases are partially or even fully ionized. Thus, the resulting ions and electrons exhibit collective effects due to interactions with electric and magnetic fields. The charge interaction, due to Coulomb forces, will obviously be greater between near neighbours. Every plasma particle has a characteristic sphere of influence with a radius defined by the Debye length, which determines a typical distance over which the impact of a charged particle's bare electric field is substantial. For distances exceeding the Debye length, the electric field of an individual charged particle is effectively shielded by the surrounding plasma. Not surprisingly, the density of plasma particles within the Debye length greatly influences the collisional properties of the plasma and determines the significance of discrete particle effects. Due to their properties the charged particles are subjugated to the so called adiabatic invariants in terms of their motion (see Fig. 3.11).

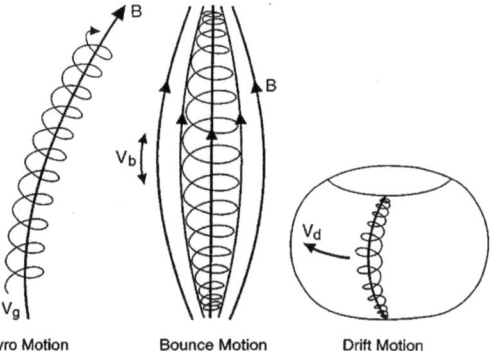

Figure 3.11: *Different types of particle motion in a magnetic field. Left figure shows a gyrating ion along a magnetic field line. In the middle plane the bounce motion of an ion, similar to the motion of charged particles between the mirror points in the north and the south hemisphere. Right figure shows the drift motion perpendicular to the magnet field lines, electrons eastwards and ions westwards. Sketch adopted from Russell [2000].*

As the charged particles are moving along the magnetic field lines, they are affected by a Lorentz force in direction perpendicular to the magnetic field and the velocity vector $v_\perp$. Therefore the particle is forced to move around the magnetic field with a frequency:

$$\omega_c = \frac{qB}{M} \qquad (3.28)$$

3.4. The Earth's Magnetic field (EMF)

where q is the electric charge of the particle, M the mass and B the magnetic flux density. Due to the velocity parallel to the magnetic field the particle performs a translation. As the magnetic field is surrendered to the particle's high velocities it experiences a force into the opposite direction if B increases. Thus, the particle slows down until it stops and the motion along the field line reverses. In this process, the perpendicular energy is

$$E_\perp = \frac{1}{2}mv_\perp^2 \qquad (3.29)$$

with $v_\perp$ as the perpendicular velocity, is proportional to the magnetic field strength, maximizing where the total energy is in perpendicular direction [Russell, 2000]. From this relation the first adiabatic invariant or also called the magnetic moment can be derived as

$$\mu = \frac{\frac{1}{2}mv_\perp^2}{B} \qquad (3.30)$$

The magnetic moment is a constant of motion and therefore has to be preserved, as the particle is gyrating along the field line (see Fig. 3.11 left figure). The conservation of the first adiabatic invariant causes the particle to reflect and bounce back and forth along the magnetic field at the point where the field lines converge. This motion is called bounce motion and in case of the Earth magnetic field this effect leads to the bouncing of the electrons and protons between the mirror points in the south and north hemisphere (see Fig.3.11 middle figure). The parallel momentum, i.e. the parallel velocity $v_\parallel$ times the mass, integrated along the field line ds, has to be conserved too. This relation is expressed by the second adiabatic invariant

$$J = \int_a^b v_\parallel ds \qquad (3.31)$$

This law shows that if the pathway along the field line shortens, the energy of the particle has to increase. In case of the particles bouncing motion in the magnetosphere at different shells, the particles have to preserve the state of equilibrium. This means, if the particles are forced to move to an inner shell e.g due to sub storms, their energy increases, the preservation of the adiabatic invariants forces the particles higher to energies.

The third important motion of plasma is the so called drift motion as can be seen in Fig. 3.9. The drift motion arises from the fact that the field line is curved and particles move parallel to the field, therefore it drifts perpendicularly to the magnetic field but in different directions depending on the charge. In our magnetosphere the combination of drift and gyration of the particles leads to the formation of the ring current. The formation of the ring current shows a significant influence on the magnetic field resulting in a sink of the magnetic field strength.

During disturbed conditions, commonly called geomagnetic storms, the Earth magnetic field fluctuates very strongly by its size and field strength up to a few days. Those fluctuations can lead to an abundance of energy in the magnetosphere which can not be balanced by the natural process e.g the polar electrojet. Thus, energy is transported and stored in the magnetospheric tail as magnetic energy. During this storing phase ($\approx$ 1h) the magnetic stress is released along the magnetic field lines into the ionosphere. Here, stored energy from the tail is converted into heat, this process is called Joule heating and is described in case of electrons by

$$Q_e = \frac{E_\perp^2 c^2}{B^2}\left[N_e \sum_j \nu_{ej} + N_i \frac{\nu_{ei}}{1+\sum \frac{\Omega_B}{\nu_{ij}}}\right] \tag{3.32}$$

where j is the current flowing in the ionospheric plasma as result of the electric field $E_\perp$, B is the magnetic field strength, ν the collision frequency with the subscripts e, i, j referring to the electrons and Ω_B as the gyrofrequency [Bauer and Lammer, 2004].

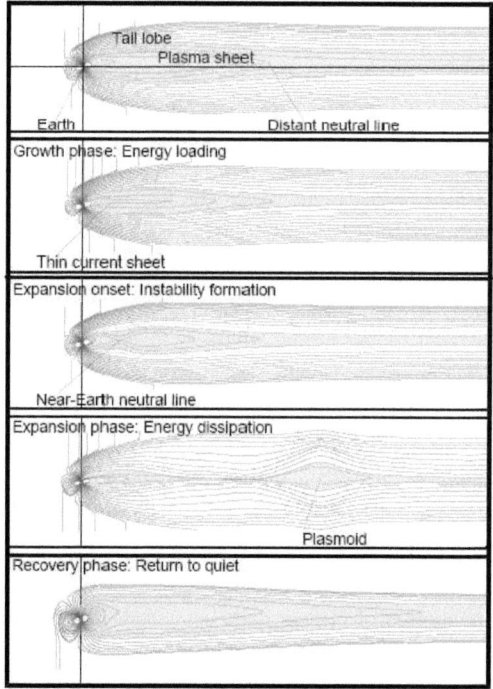

Figure 3.12: *Schematic course of a magnetospheric substorm from undisturbed conditions to recovery phase. Plot adopted from Pulkkinen [2007].*

This process of storing and dissipating of energy is called substorm, leading to phenomena like particle precipitation of several keV into polar regions, for example the

3.4. The Earth's Magnetic field (EMF)

polar lights etc. This region of increased activity is spreading westwards during the expansion phase leading also to an increase of the Birkeland currents with strengths of several Ampere km^{-2} and a higher energetic particle impact. Also the flux density of the energetic ions and electrons changes during the altering shape of the magnetosphere (see Fig. 3.12). Throughout the storage phase the tail becomes thinner and the particle flow decreases at about 6-10 Earth radii, in contrast to the substorm phase, where the changing magnetic field configuration leads to an increase in particle flux densities of several magnitudes for all kind of particles. Important to be noted is that the acceleration of the particles seems to be a local process leading to an increase in thermal energy as well as in kinetic energy. The exact processes are not well understood till now, but their effects can be observed e.g highly energetic particle precipitation and aurora lights etc.

During the substorm reconnection, lines are formed at a distance of $\approx$ 20-40 r_e. Here the antiparallel free field lines, that are located north and south of the neutral layer, are again recombined. The new connected field lines near the Earth are compressed leading to a depolarization. The connected field lines located tailwards are called Plasmoid. Charged particles in this region have energies up to several 10^{15} J and are lost as they are picked up and are carted away by the solar wind [Glassmeier and Scholer, 1991].

4. Instruments

This section introduces the instruments which provided data used in this study. Further, the quality of the measurements at certain altitude levels was investigated and compiled for different species.

4.1 Halogen Occultation Experiment (HALOE) on UARS

The Upper Atmosphere Research Satellite (UARS) was launched on September 15, 1991 aboard the Space Shuttle Discovery, starting its observation in October 1991. UARS moved on a near circular orbit at an altitude of about 585 km with an 57° orbital inclination and a period of about 96 minutes. UARS contains in total 10 scientific instruments, designed for a lifetime of 3 years but 7 of them lasted for time periods of over ten years. An overview of the instruments onboard UARS is listed in Tab. 4.4. UARS was finally switched off after 5205 days operation time on the 14th of December 2005 and has provided data spanning a long time series of over 14 years.

UARS	INSTRUMENTS
MLS	Microwave Limb Sounder
WINDII	The Wind Imaging Interferometer
HALOE	Halogen Occultation Experiment
HRDI	High Resolution Doppler Interferometer
PEM	Particle Environment Monitor
ACRIM	Active Cavity Radiometer Irradiance Monitor
SUSIM	Solar Ultraviolet Spectral Irradiance Monitor
SOLSTICE	Solar/Stellar Irradiance Comparison Experiment
ISAMS	Improved Statospheric and Mesospheric Sounder
CLAES	Cryogen Limb Array Etalon Spectrometer

Table 4.4: *Instruments onboard the Upper Atmosphere Research Satellite (UARS).*

In this study data from the Halogen Occultation Experiment (HALOE) onboard the UARS satellite were used. Due to the concerns of the deep total ozone minimum over the south pole discovered in the 1980s HALOE focused its attention on the middle atmospheric chemistry (20-50 km) and the study of stratospheric ozone and constituents and cycles leading to ozone depletion as well as chemical and dynamic properties of the mesosphere and thermosphere.

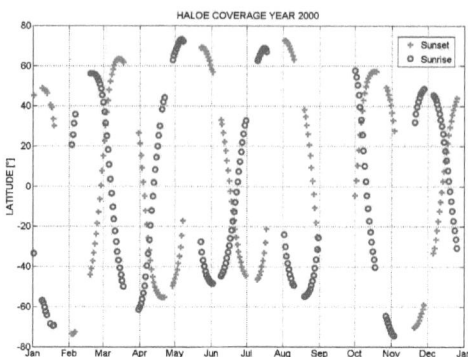

Figure 4.1: *HALOE coverage over the course of a year for sunset (SS) and sunrise (SR) measurements. In this plot the coverage of the year 2000 is shown.*

As the name suggests, the instrument uses the principles of satellite solar occultation to sound the stratosphere, mesosphere and lower thermosphere. In case of solar occultation the attenuation of the sun's radiation by the limb of the atmosphere is measured as the sun rises or sets relative to the satellite.

The theoretical approach to this technique can be seen in Fig. 4.2. If we assume $I_0(\lambda)$ is the spectrum at the highest tangent altitude with negligible atmospheric extinction and $I(\lambda, TH_1)$ is the spectrum at the tangent altitude TH_1 within the atmosphere and a certain optical depth τ, we find the relation

$$exp(-\tau(\lambda, TH_i)) = \frac{I(\lambda, TH_i)}{I_0(\lambda)} = exp\left(-\int_{Los(TH_i)} \alpha_{ext.,\lambda}(x)dx\right) \quad (4.33)$$

where we integrate along the line of sight (Los). Here $\alpha_{ext.,\lambda}(x)$ defines the total extinction coefficient at position x along the line of sight which usually is due to Rayleigh-scattering, aerosol scattering and absorption by minor constituents:

$$\alpha_{ext.,\lambda}(x) = \alpha_\lambda^{Rayleigh}(x) + \alpha_\lambda^{aerosol}(x) + \alpha_\lambda^{gases}(x) \quad (4.34)$$

Due to the different spectral characteristics of the different absorbers and scatterers the optical depth due to, for example O_3 can be extracted:

$$\tau^{ozone}(\lambda, TH_i) = \int_{Los} \underbrace{\sigma_{O_3}}_{\substack{\text{absorption}\\\text{cross}\\\text{section}}}(x) \underbrace{n}_{\substack{O_3 \text{ number}\\\text{density}}}(x)dx \quad (4.35)$$

4.1. Halogen Occultation Experiment (HALOE) on UARS

If we assume that the cross-section does not depend on x, i.e., not on temperature and/or pressure, then

$$\tau^{ozone}(\lambda, TH_i) = \sigma_{O_3} \int_{LoS(TH_i)} n(x)dx = \sigma_{O_3} c(TH_i) \tag{4.36}$$

with the column density $c(TH(i))$. The measurement provides a set of column densities integrated along the line of sight for different tangent altitudes TH_i from which vertical profiles of the species can be calculated by inversion methods [Murcray et al., 1981].

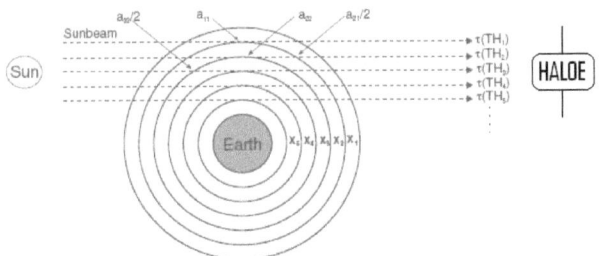

Figure 4.2: *Solar occultation geometry.*

This technique has big advantages as it is self-calibrating because each measured profile is determined by rationing the solar energy reduced by atmospheric attenuation with unattenuated solar measurements measured outside the atmosphere. But this technique of course limits the set of measurements to 15 measurements in sunrise and sunset mode every day at roughly the same latitude spaced about 25 degrees longitude apart with an instantaneous vertical field of view of 1.6 km at the Earth's limb [Russell et al., 1993]. HALOE monitors vertical distributions of HCl, HF, CH_4, and NO by gas filter correlation radiometry and H_2O, NO_2, O_3 and temperature versus pressure using CO_2 absorption by broadband filter radiometry covering selected portions of the spectral range from 2.45 μm to 10.04 μm. A latitudinal coverage from 80° south to 80° north is achieved over the period of 1 year (see Fig. 4.2) with an altitude range, depending on the species, from 15 − 130 km (see Tab. 4.5). HALOE was able to provide data from October 1991 to September 2005, thus giving the possibility to investigate the coupled dynamics and chemistry of the middle and upper atmosphere throughout a full solar cycle [Gordley et al., 1996].

The main components of HALOE can be seen in Fig. 4.3 including the sensor assembly consisting of the infrared telescope and optical mainframe, the biaxial gimbal assembly that is used for solar tracking in the azimuth and elevation directions, the sun sensor assembly, the on-gimbal electronics (GEA) and the platform electronics assembly (PEA). The ray path enters the instrument through the Cassegrain telescope consisting of a 16 cm diameter primary mirror with 96 cm focal length. Further the ray path is investigated by two different principles of instruments:

Subtype	Parameter Name	Unit	Altitude [km]	V.Res. [km]
HCL	HYDROGEN CHLORIDE	vmr	10-60	4.5
HF	HYDROGEN FLOURIDE	vmr	10-60	4.5
CH4	METHANE	vmr	15-75	4.5
NO	NITRIC OXIDE	vmr	10-130	4.5-6.5
O_3	OZONE	vmr	10-85	2-3
H_2O	WATER VAPOR	vmr	10-75	3-4
NO_2	NITROGEN DIOXIDE	vmr	10-55	2-3
TEMP	ATMOSPHERIC TEMPERATURE	K	10-130	4.5
AEXTCH4	AEROSOL EXTINCTION AT 3.46 μm	km^{-1}	10-50	2-3
AEXTCO2	AEROSOL EXTINCTION AT 2.80 μm	km^{-1}	10-50	2-3
AEXTHCL	AEROSOL EXTINCTION AT 3.40 μm	km^{-1}	10-50	2-3
AEXTHF	AEROSOL EXTINCTION AT 2.45 μm	km^{-1}	10-50	2-3

Table 4.5: *Available data products from the HALOE data server (as of 2007). Listed are subtype, parameter name, unit, altitude and vertical resolution.*

- Gas Filter Correlation Radiometer

- Broadband Filter Radiometry

The first one is the gas filter correlation radiometer. As the light enters the gas correlation chamber it is divided by optical elements into two paths. The radiation is led into different channels depending on the species NO, HCl, CH4 and HF where each channel has its own broadband optical filter and detector. The first path contains a cell filled with the specific gas of the investigated species, the second path is a vacuum path. The radiant flux in each path is then measured by infrared detectors, and the signals are analyzed. The difference in the transmission along the two paths corresponds primarily to the difference in the absorption of the gas.

The species H_2O, NO_2, CO_2, O_3 are measured by using the principles of broadband filter radiometry. Here the solar energy enters only as one path for each channel. After passing through a broadband optical filter the energy is focused on a detector. By tracking the sun, two signals are recorded, one signal during the occultation through the atmosphere, the other signal outside the atmosphere. From the ratio of the attenuated signal and the signal diffracted by the atmosphere the gas concentration can be derived.

4.1.1 HALOE Error Estimates

By operating with measurements achieved by using the gas correlation technique unfortunately one has to deal with a broad range of error sources. As the main errors arise from the changing signal coming from the sun (e.g altering solar activity etc.), the instrument was designed to be self calibrating with a calibration cycle at every measurement. Difficulties like source attenuation, refraction and source intensity were attempted to be minimized by scanning the source function (solar disk) before and after each occultation event. The second very important error source are the effects of aerosols regarding the broad band signal. These facts lead to a very low quality of the data especially at the lower stratosphere at altitudes below 25 km.

4.2. Geostationary Environmental Satellites (GOES)   43

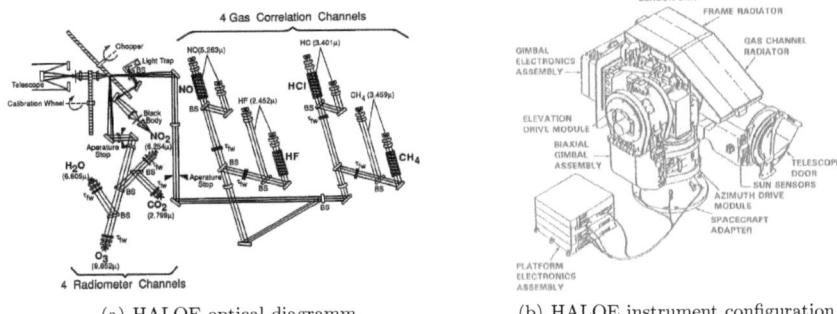

(a) HALOE optical diagramm. (b) HALOE instrument configuration.

Figure 4.3: *HALOE instrument description adopted from Russell et al. [1993].*

The 3AT Version 19 data are currently available and ready to download on the HALOE-homepage (http://haloedata.larc.nasa.gov). The data are formatted in ASCII files and contain vertical profile data of altitude versus pressure, temperature, measured quantity, and an estimated precision value. To verify the accuracy of the measurements (e.g NO, NO_2 and O_3) it is necessary to look at the quality of the data which are provided for each measurement at the data set. As the HALOE instrument was designed for investigations of the stratosphere (e.g ozone), the technical accuracy was designed to provide measurements with a high accuracy especially at altitudes between 20-45 km. To give an estimation of the total error regarding all data and each species as well as the quality, the whole data set of each species was averaged, see Fig. 4.5 - Fig. 4.9. As the error sometimes reaches a value of over 150% (private communication with J. E. Johnson NASA Goddard Space Flight Center, M. McHugh and R. E. Thompson from the GATS, Inc, responsible for HALOE data retrievals) it was necessary to evaluate the usability of the data set at certain altitudes. In case of NO the quality of the data show on average a very small error of about 2-10% at altitudes between about 20-50 km. The high values of the errors at about 80 km arise because the pressures/temperature at these altitudes are modeled (MSISE-2000) and not retrieved, therefore an increased error has to be assumed. But the accuracy to an altitude of about 110 km is still very good and can be used for further studies (private communication, R.E. Thompson, responsible for working on release for the HALOE data at GATS Inc.). A full total error estimation for the species used by this study in corresponding numbers can be seen in Tab. 4.6 and Tab. 4.7. Important to be noted is that the quality of the measurements is not significantly different for the sun rise and sun set mode. Especially in the case of HCl, hardly a difference between SS and SR seems to occur at all (see Fig. 4.9). These results are in a good agreement with the validation paper from Gordley et al. [1996]. Also recent studies of M. McHugh, GATS, comparing the HALOE measurements and the Atmospheric Chemistry Experiment ACE data, (ACE is also an occultation experiment), show quite good agreement and therefore confirm the reliability of HALOE data (private communication with M. McHugh, GATS).

Error Estimation Sun Set (SS)

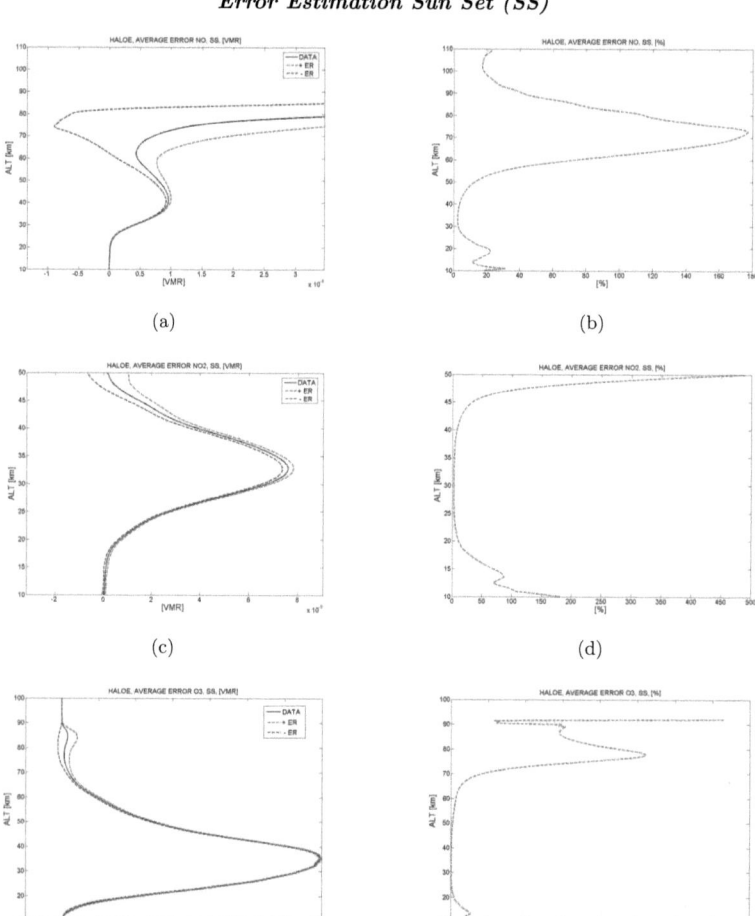

Figure 4.5: *Averaged error of all HALOE data done in sun set (SS) mode for the species NO, NO_2, O_3. Left panels (a, c, e) show the averaged species ± the averaged error in [vmr]. Right panels (b, d, f) show the calculated uncertainty in % for the same averaged data.*

4.2. Geostationary Environmental Satellites (GOES) 45

Error Estimation Sun Rise (SR)

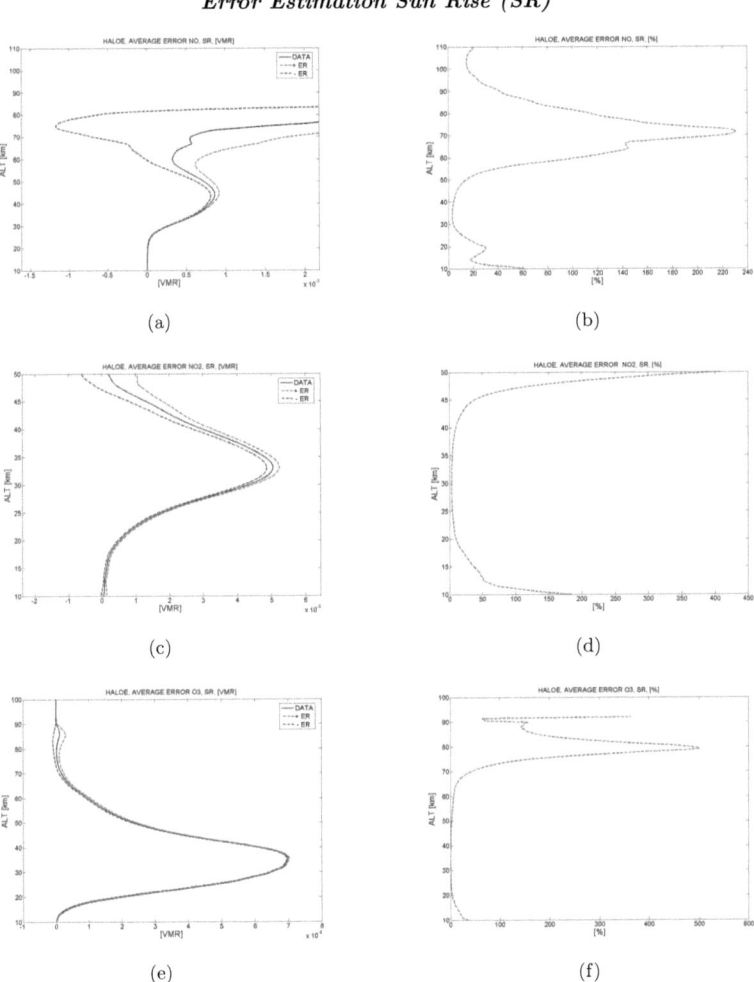

Figure 4.7: *Averaged error of all HALOE data done in sun rise (SR) mode for the species NO, NO_2, O_3. Left panels (a, c, e) show the averaged species ± the averaged error in [vmr]. Right panels (b, d, f) show the calculated uncertainty in % for the same averaged data.*

Error Estimation Sun Rise (SR) and Sun Set (SS) for HCl

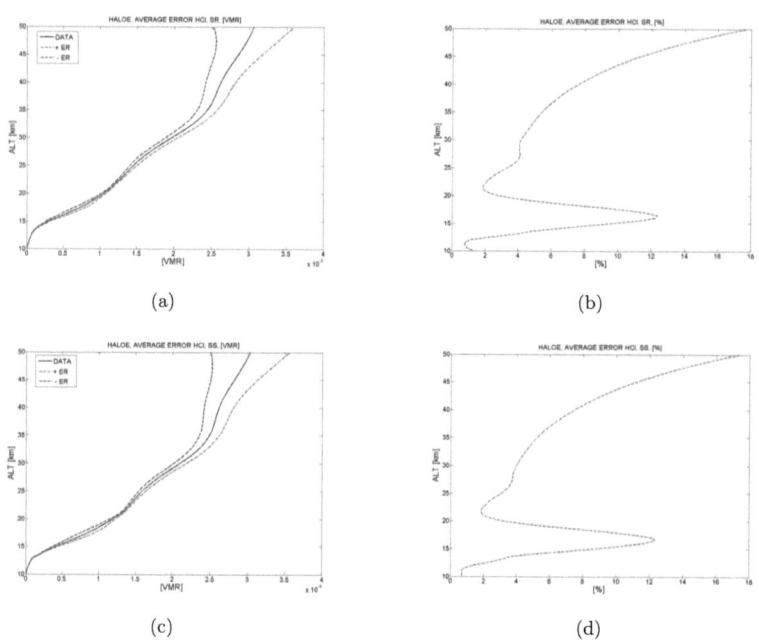

Figure 4.9: *Averaged error of all HALOE data done in sun rise (SR)mode (upper panel) and sun set (SS) mode (lower panel) for the species HCl. Left panels (a,c)show the averaged species ± the averaged error in [vmr]. Right panels (b, d) show the calculated uncertainty in % for the same averaged data.*

Alt [km]	NO[%]	NO$_2$[%]	O$_3$[%]	HCl [%]
10-20	20-30	10-130	5-20	< 12
20-45	2-10	2-8	< 5	< 10
> 50	10-180	-	10-250	< 20

Table 4.6: *Averaged error of HALOE measurements done in sun rise mode (SR) for the species NO, NO$_2$, O$_3$ and HCl, see also Fig. 4.7.*

Alt [km]	NO[%]	NO$_2$[%]	O$_3$[%]	HCl [%]
10-20	18-22	10-170	2-25	< 12
20-45	2-18	1-25	< 5	< 10
> 50	20-165	-	5-270	< 20

Table 4.7: *Averaged error of HALOE measurements done in sun set mode (SS) for the species NO, NO$_2$, O$_3$ and HCl, see also Fig. 4.5.*

4.2 Geostationary Environmental Satellites (GOES)

The Geostationary Operational Environmental Satellite (GOES) program was started by the United States back in 1974 to provide mainly meteorological data. During the years 13 satellites were launched, currently GOES 10, 11, 12 and 13 are operating. Normally 2 satellites are operating at the same time in a geostationary orbit (about 35.800 km) with 0° inclination directly over the equator. Each of them views almost a third of the Earth's surface. One focuses on North and South America and most of the Atlantic ocean, the other one on North America and the Pacific ocean. Both satellites together provide a broad range of measurements from the earth at day and night. GOES also provides important data in terms of space weather. The three main components monitored are X-rays, energetic particles and data of the Earth's magnetic field conditions. Onboard solid state detectors with pulse height discrimination measure proton, alpha particle and electron fluxes. Important for this study is the Energetic Particle Sensor (EPS). It is faced perpendicular to the spacecraft's spin axis which is approximately aligned with Earth's rotation axis. Due to its much shorter spin period (0.6 seconds) compared to its accumulation times, the EPS provides a spin averaged estimate of the local high-pitch angle particle fluxes. This study uses proton flux data of highly energetic protons with energy ranges as given in Tab. 4.8 and electron fluxes with energy range > 2 MeV. Data with a sampling rate of 1 day in units of $ions/cm^2/s/sr$ were used. For investigations of highly energetic solar particle interaction with the earth middle atmosphere the proton channels G3, G4, G5 were used. Data and further information is available on the Space Physics Interactive Data

Resource (SPIDR) homepage at http://spidr.ngdc.noaa.gov.

G_1	G_2	G_3	G_4	G_5	G_6	G_7
0.8-4 MeV	4-9 MeV	9-15 MeV	15-40 MeV	40-80 MeV	80-165 MeV	165-500 MeV

Table 4.8: *GOES proton channels available from the SPIDR.*

Figure 4.10: *Geostationary orbit of the GOES satellites.*

4.3 Polar Operational Environmental Satellite (POES)

The Polar Operational Environmental Satellite (POES) Program is a cooperative effort between NASA and the National Oceanic and Atmospheric Administration (NOAA), the United Kingdom (UK), and France that started in the early 1960s with mainly meteorological purpose. Currently, the POES mission is composed of two polar orbiting satellites. Operating as a pair, these satellites primarily provide data used for long-range weather forecasting ensuring that infrared and non-visible data for any region of the Earth are no more than six hours old (POES hompage: http://goespoes.gsfc.nasa.gov). The spacecrafts operate in a circular, near polar orbit of about 833 km with an inclination of approximately 98.6° to the equatorial plane. POES orbits sun-synchronously with an averaged period of about 101 minutes. In contrast to the GOES satellites, which are focused on near term data, the POES project provides full global data for short, medium and long range forecast models, climate modeling and environmental studies. The instrument onboard which is important for this study is the Medium Energy Proton and Electron Detector (MEPED) as a part of the Space Environment Monitor (SEM-2). The instrument is focussed on measurements of electrons in the energy range from 30 keV up to 2.5 MeV by using 2 detectors [Wissing et al., 2008]. One is looking backwards along the satellite trajectory the other one has a viewing direction radially outwards. For the purpose of this study data from the 0° detector were taken from the currently orbiting satellites NOAA-15 and NOOA-16. The electron fluxes measured from the 0° are the particles which are entering the atmosphere and are important for the investigations of EEPs. As the spacecraft is moving at high latitudes it roughly looks parallel to the magnetic field lines, therefore POES detects precipitating

4.3. Polar Operational Environmental Satellite (POES)

particles, in contrast to the 90° detector which mostly measures those particles that are mirrored back before hitting the atmosphere. Furthermore separate measurements of the morning and evening sectors have been made showing dependence of precipitating particles (≈50 keV) and the local time [Wissing et al., 2008]. The final data set used in this study was obtained from Jan Maik Wissing, University of Osnabrück, Germany. This data set contains the corrected electron flux in counts/s and is separated into three channels (see Tab. (4.9), e1_e2, e2_e3 and e3 containing 3 different energy ranges from 30 keV up to 2.5 MeV with a geometrical factor of the detector of 0.01 cm^2 sr.

e1_e2	e2_e3	e3
30-100 keV	100-300 keV	> 300 keV - max.2.5 MeV

Table 4.9: *POES data obtained from University of Osnabrück.*

Figure 4.11: *Polar orbit of the NOAA POES satellites.*

4.4 Ground-based measurements: Ap-index

The Ap-index is a well known proxy for the activity of the geomagnetic field and thus connected to the solar interaction with the Earth's magnetic field and consequential particle impacts e.g. solar cycle, semiannual variation, solar rotation, energetic electron precipitation (EEP) and solar proton events (SPE) [McPherron, 1999, Menvielle and Berthelier, 1991]. The daily Ap index is derived from the 3-hour Kp index which is monitored by ground-based magnetic observatories recording the three magnetic field components. The global Kp index is obtained as the mean value of the disturbance levels in the two horizontal field components, observed at 13 selected subauroral stations [Menvielle and Berthelier, 1991]. The scale is O to 9 expressed in thirds of a unit, e.g. 5- is 4 2/3, 5 is 5 and 5+ is 5 1/3. This planetary index is designed to measure solar particle radiation by its magnetic effects (see also NOAA hompage). The 3-hourly ap (equivalent range) index is derived from the Kp index as follows from Tab. 4.10. Detailed investigations of the Ap-index and how it is connected to electron flux data obtained from GOES and POES is shown in chapter 7.1.

Kp	0o	0+	1-	1o	1+	2-	2o	2+	3-	3o	3+	4-	4o	4+
ap	0	2	3	4	5	6	7	9	12	15	18	22	27	32
Kp	5-	5o	5+	6-	6o	6+	7-	7o	7+	8-	8o	8+	9-	9o
ap	39	48	56	67	80	94	111	132	154	179	207	236	300	400

Table 4.10: *The 3 hourly Ap-index derived from the Kp-index.*

5. Solar Proton Events (SPE)

The occurrence of large solar eruptions is often followed by so called Solar Proton events (SPE) which have been scientifically observed since the early 1960s (see also section. 3.2.1). This chapter introduces the chemical impact of highly energetic particles originated at the sun and how they are affecting the atmospheric chemistry. Results from investigations of the HALOE data set, covering the years 1991 - 2005, regarding SPE impacts are shown. This section focuses on the species NO, NO_2 and HCl and their response to SPEs, which are involved in catalytic processes leading to ozone depletion e.g. the NO_x cycle. Also, a possible effect on temperature during solar proton events is investigated.

5.1 Properties of SPEs

During the time period 1991 - 2005, spanning a part of the solar cycle 22 and the whole solar cycle 23, at least two very strong SPEs occurred during solar maximum and got reasonably famous because they are quite well covered by observations (e.g. the papers of Jackman et al. [2001, 2005b] and many others). One of this famous events happened during the days from the 10th to 15th of July 2000 and is also called the 'Bastille - Event' (see Fig. 5.1) the other one happened during 29th and 31th of October 2003 and was followed by a second very active phase during the 3rd and 5th of November 2003 (see Fig. 5.2) and is called the 'Halloween - Event'. During the phase of the Bastille event, the HALOE spacecraft was measuring at polar regions in the northern hemisphere, enabling atmospheric measurements covering the whole range of the event. In case of the Halloween event the spacecraft was not able to collect data of the whole event, thus this study focuses on the event in July 2000 as it is representative for large SPEs.

The Bastille event started a few days in advance of the July 14, 2000, as smaller plasma ejections with shock speed velocities of about 700 km s^{-1} were ejected from the solar disk. During the 10th to 15th of July 2000, 3 X-class flares and two halo CMEs could be observed by the Solar and Heliospheric Observatory (SOHO) spacecraft [Bombardieri et al., 2006]. Those phenomena caused major disturbances to the interplanetary magnetic field. These disturbances affecting the magnetic field were observed by several spacecrafts and also by in situ measurements on earth for example the Ap-index (see Fig. 5.1). The daily Ap-index, and thus a response of the geomagnetic field, starts increasing by a value of about 25 on the 10th of July, reaching its peak on the 15th of July with a value of over 150. In case of the Halloween event also an Ap-index of

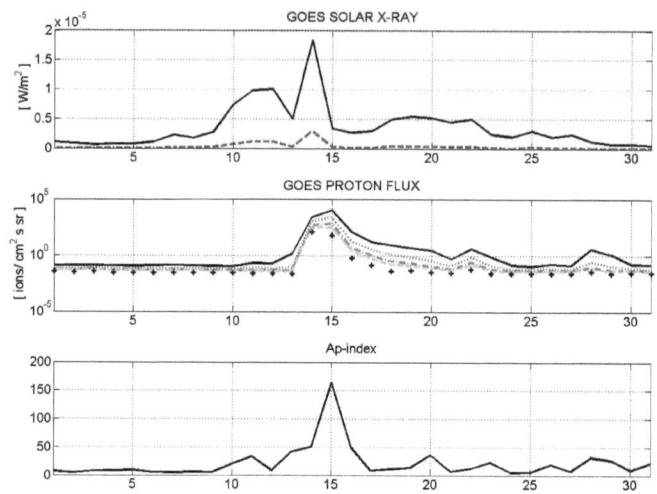

Figure 5.1: *Measurements from the GOES spacecraft for July 2000, sampling rate 1 day. Upper panel: Solar X-rays 1-8 Å (solid line) and solar X-rays 0.5-4 Å (dashed line). Middle panel: Proton measurements from 10 to 100 MeV. Protons >10 MeV (black solid line), > 30 (blue dotted line), > 50 (red dashed dotted line), > 60 (green dashed line) and + icons show protons > 100 MeV. Lower panel: Daily Ap-index. Data obtained from the Space Physics Interactive Data Resource (SPIDR), http://spidr.ngdc.noaa.gov/spidr/home.do.*

approximately 200 has been observed during the main event lasting from the 29th to 31st of October 2003 (see Fig. 5.2). The primary source of the SPE on the July 14, 2000 was located near the solar meridian N 22°, W 22°, started at 10:03 UT, reached its maximum at 10:24 UT and finally ended at 10:46 UT. During this time intense long duration type III bursts from microwave to hectometric wavelengths were detected associated with electron acceleration deep in the solar corona [Reiner et al., 2001]. Also the occurrence of soft to hard X-rays was observed by the GOES 10 detectors (see Fig. 5.1). Data show an increase of middle to soft X-rays (1 - 8 Å) starting with the 10th of July, the maximum is reached on the 14th of July, reaching peak values of 1.841×10^{-5} W/m^2. A significant increase in middle to hard X-rays (0.5 - 4 Å) can only be observed after the big flare on the 14th of July reaching a peak value of 2.968×10^{-6} W/m^2. The acceleration of charged particles due to solar events is induced by several mechanisms. These include processes like diffusive shock acceleration at the bow shock of a CME, resonant wave-particle interactions initiated by magneto hydrodynamic (MHD)

5.2. Effects on the middle atmosphere

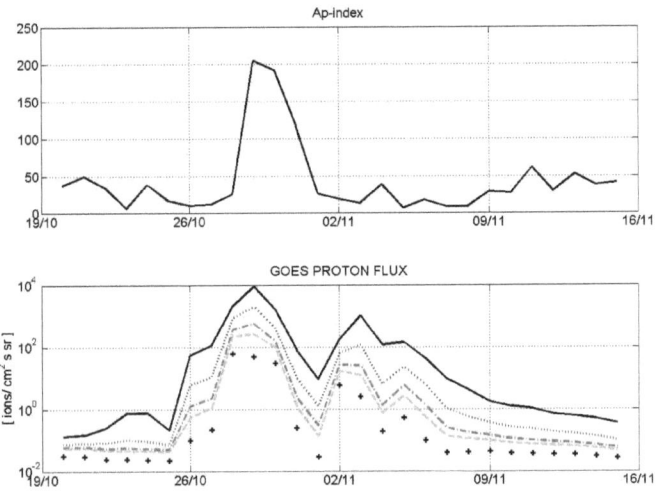

Figure 5.2: *Measurements from the GOES spacecraft for mid-October to mid-November 2003, sampling rate 1 day. In case of the time range October to November 2003 solar X-ray data are not available. Upper panel: Daily Ap-index. Lower panel: Proton measurements from 10 to 100 MeV. Protons >10 MeV (black solid line), > 30 (blue dotted line), > 50 (red dashed dotted line), > 60 (green dashed line) and + icons show protons > 100 MeV. Data obtained from the Space Physics Interactive Data Resource (SPIDR), http://spidr.ngdc.noaa.gov/spidr/home.do.*

turbulence and direct acceleration by DC electric field currents in neutral current fields. These processes are discussed by e.g. Bombardieri et al. [2006] and Miller et al. [1997]. In case of protons, a sudden increase of fluxes with energies ranging from keV up to relativistic protons (430-745 MeV) could be detected by the GOES detectors. GOES 10 daily averaged proton data show a maximum in flux on the July 15, 2000 at several channels from 10 to over 100 MeV (see Fig. 5.1). The maximum of proton flux between 10 and 100 MeV during the Bastille event as well as during the Halloween event is shown in Tab. 5.11. To get an impression of the massive increase of proton flux caused by SPEs, also values of the proton flux from the February 21, 2008, is shown, where no SPE has happened.

Three days after the Bastille event the Ap-index turns back to undisturbed conditions, the lower energetic proton fluxes still stay slightly increased until the end of the month July, as well as the soft X-ray.

SPE	p > 10 MeV	p > 30 MeV	p > 50 MeV	p > 60 MeV	p > 100 MeV
Bastille Event	9840	2344	647	301	55
Halloween Event	8941	2034	577	261	49
February 21, 2008	0,20	0,11	0,08	0,08	0,04

Table 5.11: *GOES daily averaged data of several proton channels covering energy ranges from > 10 to > 100 MeV in ions/ $cm^{-2}-s-sr$. Shown are proton data of the Bastille event in July 2000, the Halloween event in October 2003 and data of February 21, 2008, during a time where no SPE occurred.*

5.2 Effects on the middle atmosphere

5.2.1 Nitrogen compounds

The abundance of nitrogen oxides and related nitrogen compounds throughout the middle atmosphere is mainly affected by two main processes. On the one hand due to oxidation from nitrous oxide N_2O and on the other hand due to ionization from molecular nitrogen N_2 by highly energetic particles. Other sources, becoming a more and more serious topic in times of the anthropogenic air pollution, are aircraft engines, cultivated soils, animal waste, biomass burning, injecting nitrogen oxides in the lower stratosphere. But there are of course natural sources which play a major role in the atmospheric equilibrium like the oceans and tropical soils, investigated and published by the WMO report 2006.

The important species NO and NO_2 are in photochemical balance during day time. A permanent and fast conversion of NO into NO_2 by interaction with ozone is leading to large amounts of catalytically destroyed ozone. The cycle is limited by the transformation of nitrogen oxides into less reactive nitrogen compounds. This happens for example during night time condition (see Eq. 5.37). Here, NO_2 is converted, as it is consuming ozone, into NO_3, and is stopped because the photolytic destruction ceases during night. Further, the production of bigger and more stable molecules is engaged as it reacts with NO_2 to form dinitrogen pentoxide N_2O_5 or nitric acid HNO_3.

$$\begin{aligned} NO_2 + O_3 &\rightarrow NO_3 + O_2 \\ NO_3 + h\nu &\rightarrow NO + O_2 \\ &\rightarrow NO_2 + O \\ NO_3 + NO_2 + M &\rightleftharpoons N_2O_5 + M \end{aligned} \quad (5.37)$$

After sunrise, the stable molecules are once again photolyzed depending on several factors e.g. solar zenith angle, albedo and ambient temperature, leading to steady state conditions of the NO/NO_2 ratio. This release of NO_x from N_2O_5 formed at night is partially responsible for the growth during the morning hours.

$$\begin{aligned} N_2O_5 + h\nu &\rightarrow NO_3 + NO_2 \\ NO_3 + h\nu &\rightarrow NO + O_2 \\ &\rightarrow NO_2 + O \end{aligned} \quad (5.38)$$

5.2. Effects on the middle atmosphere

In the thermosphere (≥ 100 km) atomic nitrogen is the result of dissociation processes of N_2 engaged by solar radiation and energetic particle precipitation, thus this region is strongly connected to the solar variability. Nitrogen is primarily found in the form of NO and N. It is produced because of the absence of ozone at this altitude, thus the conversion of NO to NO_2 is not possible. Here, the concentration of atomic oxygen is high and hence the major reaction of nitrogen results in an abundance of NO. The substantial production of N and NO occurs due to ionic processes and N_2 photolysis in thermospheric regions. NO is transported downward to the mesosphere where it is photolyzed and again reformed into stable N_2 by recombination processes between N and NO. Due to the short lifetime of nitrogen, the steady state conditions can only be assumed for atomic nitrogen in ground state and excited state, hence Eq. 5.41 and Eq. 5.42 are the main processes for the formation of NO in the mesosphere. The coupling of the thermosphere and the upper mesosphere is very weak. But during wintertime at polar regions, and thus polar night conditions, photolysis of NO_x can not occur. Thus the species keep stable and are not destroyed because of the missing solar radiation. These long lived species are now transported by the general circulation from the thermosphere to lower altitudes and lead to an increase of NO_x and other species in the mesosphere and stratosphere [Brasseur and Solomon, 2005].

All these 'natural' processes throughout the thermosphere, mesosphere and stratosphere, lead to the typical structure of NO_x volume mixing ratios versus altitude, shown in Fig. 5.3 (upper panel). Here, at an average latitude of about $\sim 70°$ daily average values over 5 different days in 5 different years in the month April during low solar activity were averaged. The values are strongly depending on latitude, season and solar activity, especially at regions exceeding approximately 80 km. A seasonal variation of NO_2 calculated by averaging NO_2 profiles covering the years 1991-2005 is shown in Fig.5.3 (lower panel).

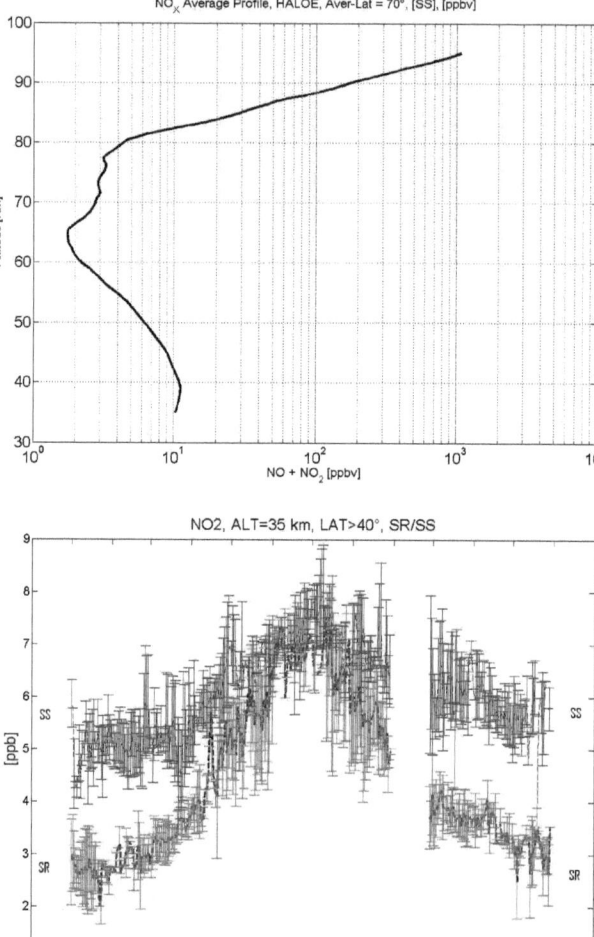

Figure 5.3: *Upper panel:* Average NO_x profile versus altitude for polar region (latitude ≈ 70°) in ppbv derived from HALOE data. *Lower panel:* Daily averaged seasonal profile of NO_2 derived from HALOE data at 35 km altitude, in sun rise (SR) and suns set (SS) mode. Profile derived from data covering the years 1991 to 2005 at latitudes > 40°.

5.2. Effects on the middle atmosphere

SPEs lead to an increase of highly energetic particles which can affect atmospheric species deep down to the middle atmosphere and thus lead to many chemical interactions with species of the neutral atmosphere. Hence, the production of HO_x (H, OH, HO_2), NO_y (NO, NO_2, NO_3, N_2O_5, HNO_3, HO_2NO_2, HONO, $ClONO_2$, $ClNO_2$, $BrONO_2$) and even Mg^+ and Mg, as shown by recent studies by Scharringhausen [2008], increases directly or due to a photochemical sequence [Vitt and Jackman, 1996]. A very important chemical process is the production of odd nitrogen in the middle atmosphere because of its great influence on the ozone depletion in the mesosphere and even in the stratosphere. Impacts of protons dissipate their energy in creation of ion pairs within the neutral atmosphere. Ion pairs are created as a proton removes an electron from a neutral atom or molecule. This electron is the so called secondary electron. This reaction leaves a positive ion. The resulting freed electron adopts energy of the proton and causes further ionization in the atmosphere [Jackman et al., 2005b]. This process is leading to several chemical reactions:

$$e^\star + N_2 \rightarrow N_2^+ + 2e$$

$$e^\star + N_2 \rightarrow N + N^+ + 2e$$

$$e^\star + N_2 \rightarrow N + N + e \qquad (5.39)$$

$$e^\star + O_2 \rightarrow O_2^+ + 2e$$

$$e^\star + O_2 \rightarrow O + O^+ + 2e$$

These are subsequently followed by interchange and recombination reactions producing atomic nitrogen [Rusch et al., 1981].

$$N_2^+ + O \rightarrow NO^+ + N$$

$$N_2^+ + e \rightarrow N + N$$

$$O^+ + N_2 \rightarrow NO^+ + N \qquad (5.40)$$

$$NO^+ + e \rightarrow N + O$$

$$N^+ + O_2 \rightarrow O^+ + NO$$

$$\rightarrow O_2^+ + N$$

Finally nitric oxide is produced subsequently by the reaction of ground state [$N(^4S)$] level or the exited [$N(^2D)$] level with O_2.

$$N(^4S) + O_2 \rightarrow NO + O \tag{5.41}$$

$$N(^2D) + O_2 \rightarrow NO + O \tag{5.42}$$

The reaction Eq. 5.42 is the faster one under middle atmospheric conditions. Following Porter et al. [1976]. 1.25 N atoms are produced per ion pair and can be separated in production of N atoms by $\sim 45\%$ for ground state and $\sim 55\%$ for excited state nitrogen atoms. The mutual destruction of odd nitrogen occurs through

$$N(^4S) + NO \rightarrow N_2 + O \tag{5.43}$$

$$N(^2D) + NO \rightarrow N_2 + O \tag{5.44}$$

In a publication by Jackman et al. [2001] the production of NO_y molecules was modeled focusing on the last 4 great SPEs. In case of the SPE in July 2000 a production of 3.5×10^{33} molecules of NO_y was shown. The investigation of NO_x production is very important because of its destroying effect on ozone in the mesosphere and stratosphere. HO_x molecules are short lived but cause a significant ozone depletion in the middle atmosphere whereas the longer lived NO_x compounds are transported downward into the stratosphere during polar winter and lead to an additional catalytic O_3 loss by the reactions [Brasseur and Solomon, 2005]:

$$NO + O_3 \rightarrow NO_2 + O_2 \tag{5.45}$$

$$NO_2 + O(^3P) \rightarrow NO + O_2 \tag{5.46}$$

This catalytic NO_x cycle is most effective between 30 and 40 km with a peak rate of 10^6 molecules cm^{-3} s^{-1} leading to ozone depletion [Lary, 1997].
The most effective catalytic cycle causing ozone depletion in the upper stratosphere is shown in Eq. 5.47- 5.49. This cycle also reaches a peak rate of 10^6 molecules cm^{-3} s^{-1} but at altitudes of about 50 to 70 km [Lary, 1997]. In this cycle ozone is not destroyed directly but this cycle expends a great amount of atomic oxygen and this is not further available for building ozone.

$$H + O_2 \rightarrow HO_2 \tag{5.47}$$

$$HO_2 + O(^3P) \rightarrow OH + O_2 \tag{5.48}$$

$$OH + O(^3P) \rightarrow H + O_2 \tag{5.49}$$

5.3 Results (SPE)

5.3.1 Long term observation of NO_x

After the launch of the UARS spacecraft on the September 15, 1991, HALOE started its measurements in sun rise (SR) and sunset (SS) mode (see also chapter. 3.4.1) in the following month. Data of NO V19 3AT vmr are available from the October 11, 1991 to the September 10, 2005 in SS and SR mode, each mode containing of approximately 15 measurements a day. Fig. 5.4 shows the whole set of NO_x data in SS mode separated in northern and southern hemisphere. NO and NO_2 data were daily averaged and then summed up to calculate NO_x volume mixing ratios for altitudes from 35 to 95 km. The general feature of the NO_x is fairly constant and comparable to the 'standard' profile shown in Fig. 5.3.

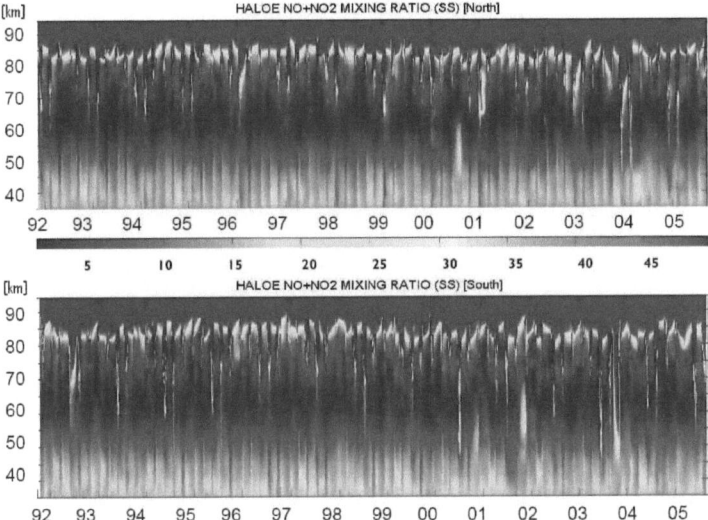

Figure 5.4: *Contour plot of daily averaged values (interpolated) of NO and NO_2 mixing ratio [ppbv] covering the years from 1991 to 2005. Altitudes from 35 km to 95 km sunset mode (SS) northern hemisphere averaged over all latitudes from $0°$ to $90°$ (upper panel) and southern hemisphere averaged over all latitudes from $0°$ to $-90°$ (lower panel).*

At altitudes between 35 and 50 km NO_2 is dominating the NO_x budget. At altitudes higher than 50 km a sink of NO can be seen and at altitudes over 80 km a sudden increase of NO arises due to the thermospheric coupling. At this altitude NO varies very strongly due to altering solar radiation, particle precipitation and seasonal variations.

The complete data set is very long, therefore effects of smaller particle precipitation can not be observed very well by contour plots, some sudden increases at low altitudes are sometimes caused by the sudden latitudinal changes of the spacecraft. Hence, the graphical interpolation function of the programming language sometimes tends to interpolate over time ranges where no data are available and shows sudden increases which are not real. Thus, suspicious events have been checked by looking into the single measurements of the data set. In Fig. 5.4 (lower panel) the Halloween event can be seen with a strong signal reaching down to altitudes of lower than 50 km in the Southern hemisphere.

A better resolution of strong effects caused by solar particle interaction can be seen in Fig. 5.5. A very strong signal of NO increase associated with the Bastille event in 2000 arises down to altitudes beyond 50 km and can be observed in SR mode on the Northern hemisphere. Further the smaller SPE during 7th to 8th November 2004 and the 16th to 17th January 2005 can be observed. A significant increase of NO down to altitudes of approximately 65 km can be seen very well.

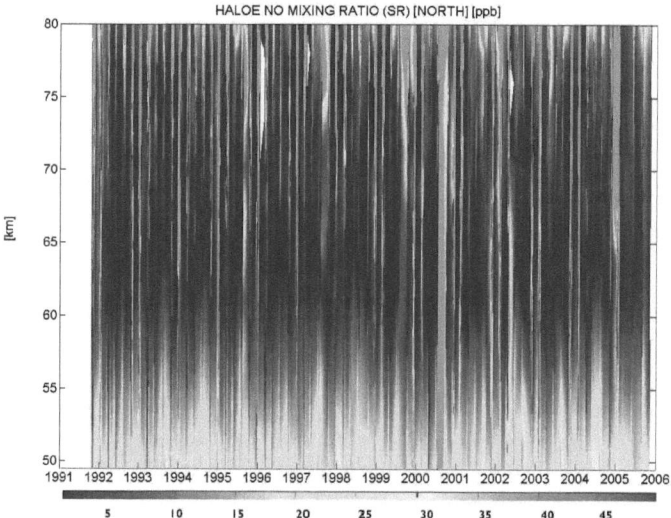

Figure 5.5: *Contour plot of daily average values (interpolated) of NO covering the years from 1991 to 2005 from 50 to 80 km altitude in sun rise mode of the northern hemisphere. The very strong increase of NO_x during the July 2000 event and the very active phase in January 2005 can be seen.*

Fig. 5.6 - Fig. 5.9 show daily averaged data for the years 1991-2005 for SS and SR mode from latitudes > 40 ° for some selected altitudes. The seasonal variations due to

5.3. Results (SPE)

the general circulation and the transport of NO from the thermosphere can be observed very well, especially at the altitudes of 80 km and 70 km. In case of the northern hemisphere the NO vmr are increased during the winter months November, December, January and February (see Fig. 5.6 and Fig. 5.8). In the southern hemisphere the winter lasts for the months May, June, July and April, the increased NO vmr can be seen in Fig. 5.7 and Fig. 5.9. A strong variation of winter time NO from year to year can also be observed. These variations are caused by changes of energetic electron precipitation, and are investigated more detailed in chapter. 6.1.4

In Fig. 5.6 the effects caused by the Bastille event can be seen very strongly at altitudes from 70 - 50 km. Also the second phase of the Halloween event in early November shows a significant NO increase. Fig. 5.9 shows a very strong increase of the NO, induced by the Halloween event during October 2003. Investigation of long term NO_2 (long term series not shown) showed peak values of 1.9 ppb, compared to undisturbed conditions an increase of 1.5 ppb at 50 km during the Bastille event, and for the Halloween event a peak value of 1.3 ppb which means an increase of 0.9 ppb compared to undisturbed conditions. All other peaks could not be referred to a special solar proton event, other SPEs have not been observed by HALOE due to its low latitudinal coverage.

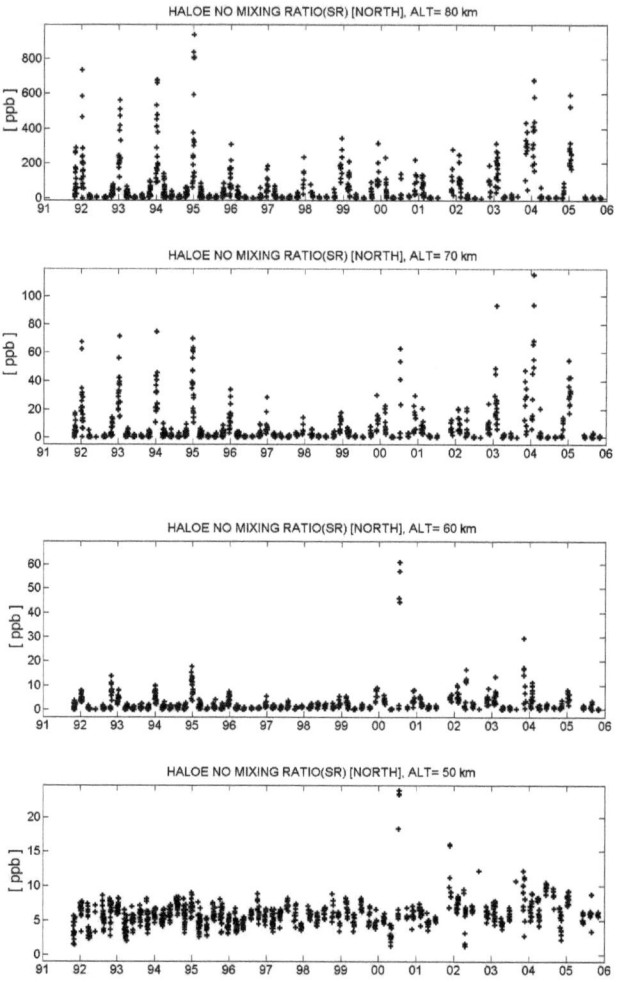

Figure 5.6: *HALOE NO vmr daily averaged values at several altitudes for latitudes > 40° north covering the years 1991 to 2005 in sun rise mode (SR).*

5.3. Results (SPE)

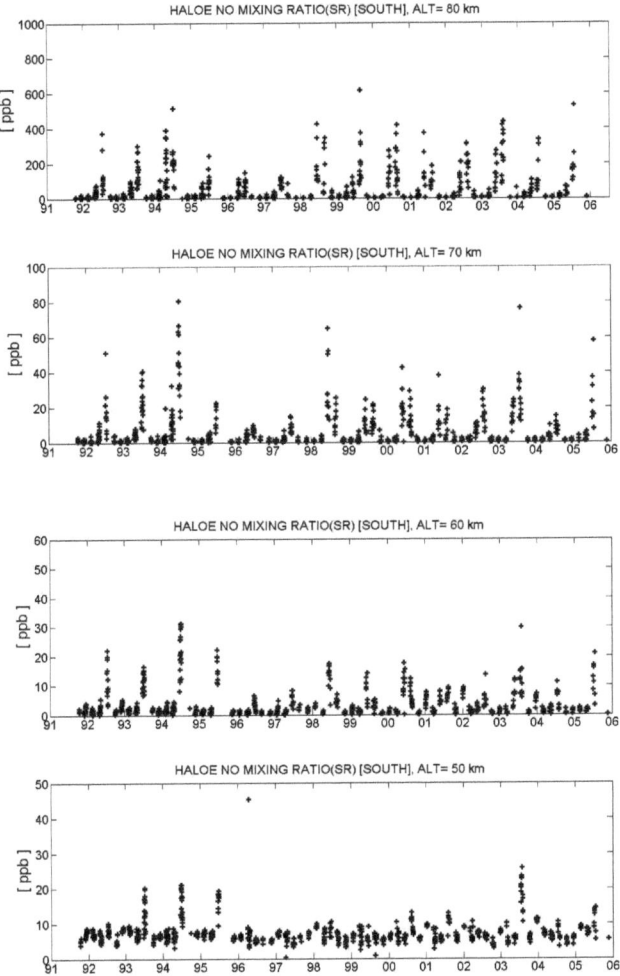

Figure 5.7: *HALOE NO vmr daily averaged values at several altitudes for latitudes > 40° south covering the years 1991 to 2005 in sun rise mode (SR).*

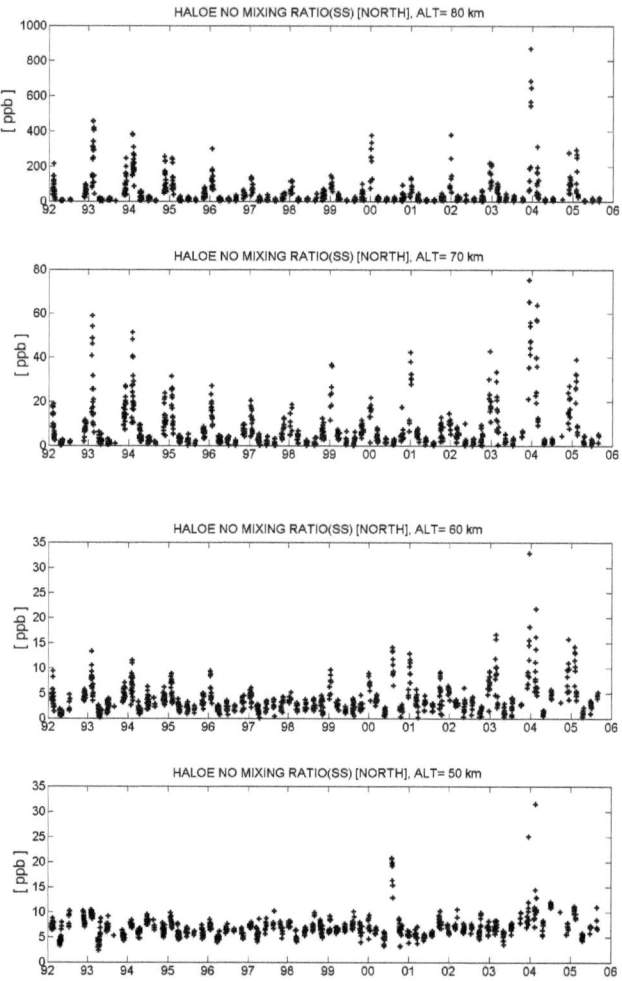

Figure 5.8: *HALOE NO vmr daily averaged values at several altitudes for latitudes $> 40°$ north covering the years 1991 to 2005 in sun set mode (SS).*

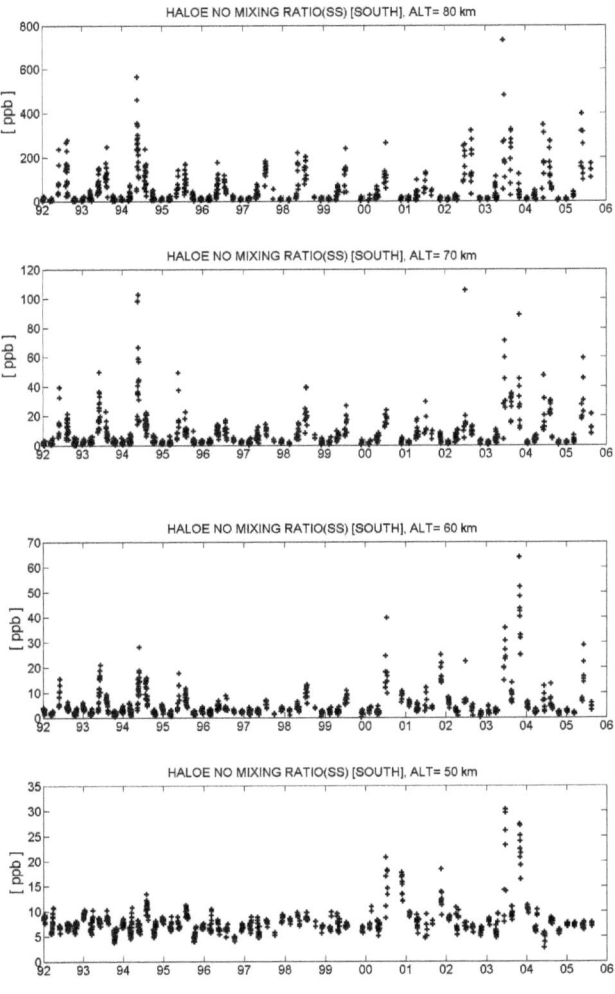

Figure 5.9: *HALOE NO vmr daily averaged values at several altitudes for latitudes > 40° south covering the years 1991 to 2005 in sun set mode (SS).*

Long term studies have also been carried out for ozone, but do not show significant results at a first glance, therefore the data series are not shown, but the strong coupling of ozone and NO during solar proton events will be discussed later in section 5.3.3. Investigations of H_2O showed no significant results during the July 2000 SPE and a direct connection could not be found.

5.3.2 Long time observation of HCl

In terms of HCl, measured by HALOE between altitudes from 10 to approximately 60 km, some interesting features were observed. HCl is an important and quite stable reservoir for chlorine. It is indirectly involved in production of odd hydrogen (HO_x = H + OH + HO_2) which is supposed to lead to a major ozone depletion in the upper stratosphere and lower mesosphere [Rusch et al., 1981]. A reference profile of HCl from 20 to 60 km is shown in Fig. 5.10. It is known that odd hydrogen removes atomic chlorine from the reservoir HCl via the reaction

$$OH + HCl \rightarrow H_2O + Cl \qquad (5.50)$$

The freed atomic chlorine can further react with ozone by the reaction

$$Cl + O_3 \rightarrow ClO + O_2 \qquad (5.51)$$

and further can give rise to enhanced HOCl values by reaction

$$ClO + HO_2 \rightarrow HOCl + O_2 \qquad (5.52)$$

This set of reactions leads on the one hand to a direct ozone depletion by Eq. 5.51 and on the other hand SPEs lead to enhanced HO_x abundance, which also has a major influence on ozone throughout the middle atmosphere [von Clarmann et al., 2005], but under normal conditions, this release of chlorine from HCl is very slow in the upper stratosphere and lower mesosphere.

The long time series of HCl data in Fig. 5.11 shows two interesting effects. At first, from the year 1991 to 1997 a continual increase of HCl at altitudes between 30 and 60 km could be observed. Since it was known that chlorofluorocarbons CFCs (HCl is formed from reactions with chlorofluorocarbons CFCs) from natural and anthropogenic sources are involved in ozone destruction, the 'Montreal protocol' for the protection of our atmospheric system was inured back in 1987. Laws were introduced to decrease the anthropogenic chlorine production as it directly or indirectly promotes ozone depleting processes. As can be seen in Fig. 5.11, the increase of HCl at altitudes over 40 km stops in the years 96 and 97 remaining nearly constantly throughout the years 98 to 2002, and seems to even decrease in the years 2003 and 2004.

The second effect can be seen during the SPE in July 2000 (see Fig. 5.11). Exactly on the 15th of July 2000 a decrease of HCl of $\approx$ 0.5 ppb can be observed within the long time series between 50 and 40 km altitude. Due to the Eq. 5.50, a sink of HCl should be the logical effect during a SPE, as the abundance of HO_x compounds, produced during the SPE, reacts with the chlorine reservoirs. Evidence for this effect is discussed in detail in section 5.3.5 and model studies regarding the chlorine activation can be found in section. 6.1.3.

5.3. Results (SPE)

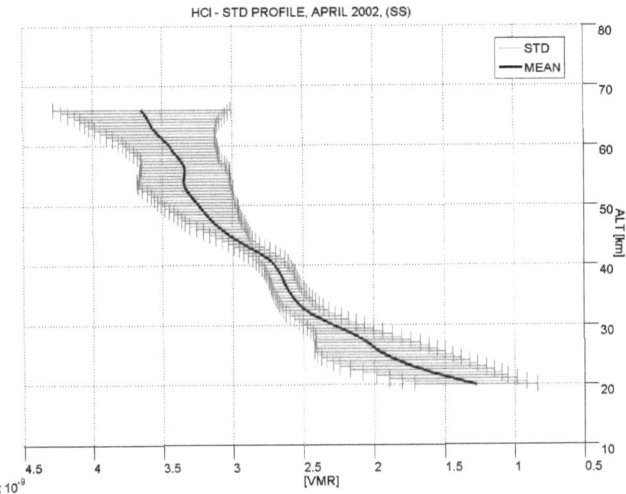

Figure 5.10: *HCl profile, derived from monthly averaged HALOE data in April 2002 at latitudes from 19° to 79° North measured in suns set mode.*

5.3.3 Bastille Event - NO_x

Investigations of the long term time series in section 5.3.1 showed effects down to altitudes of 50 km that are correlated with well known SPEs during these years. Thus, more detailed investigations were carried out for the species NO, NO_2, O_3 and HCl. As the highly energetic protons are known to penetrate the earth atmosphere along the opened magnetic field lines, only measurements at geomagnetic latitudes greater than 60° were taken into account. Hence, HALOE data during the Bastille event fulfill these terms of conditions and are recorded during the whole event.

In Fig. 5.12 (upper panel) NO measurements throughout the 11th to 18th July at geomagnetic latitudes greater than 60° in sun rise mode are shown, below the GOES proton channels with energy ranges 9-15 MeV (black dashed line) and 15-40 MeV (red dashed line) are plotted for the same time range. The effects of the highly energetic protons can be seen starting on the 15th of July, approximately about 20 hours after the first big ejections from the sun, as particles need time to travel the distance to the earth. A strong increase down to altitudes of about 45 km can be seen lasting for two days, till the 17th of July. Although the proton flux returns to normal values again on the 16th of July, the NO values are still increased and start to turn slowly back to undisturbed conditions from the bottom up to the mesosphere. In Fig. 5.14 NO values are shown in detail for some selected altitudes at 80, 70, 60, 50, 45 and 40 km. A clear increase of NO during the SPE and till the 18th of July at altitudes below 80

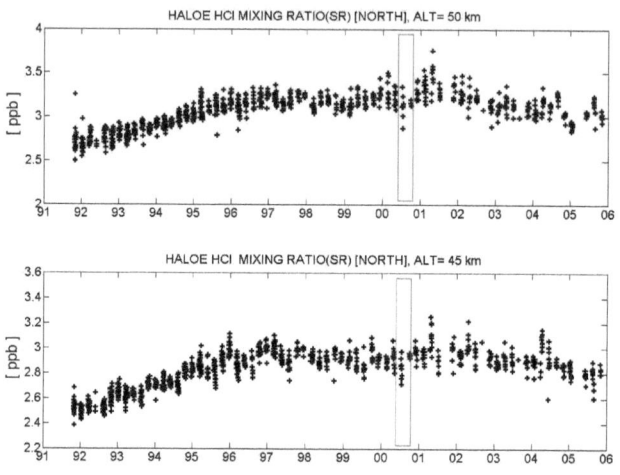

Figure 5.11: *HALOE HCl vmr daily averaged values at 50 and 45 km altitude for latitudes > 40° north covering the years 1991 to 2005 in sun rise mode (SR).*

km can be seen, as the recovery process starts from the upper regions. Calculation of the daily averaged NO values during the event showed up to 40 times higher NO values at altitudes 65 - 75 km in the mesosphere, decreasing with altitude to a factor of ≈ 5 at 45 - 50 km (see Fig. 5.13(a)). The total differences in volume mixing ratio of daily averaged NO during the disturbed and undisturbed conditions show an increase of up to 300 ppb during the SPE at altitudes of ≈ 80 km (see Fig. 5.13(b)).

The increase of the NO_2 was investigated separately (see Fig. 5.15). NO_2 data were only provided up to altitudes of 50 km. At altitudes of 50 km ≈ 20 higher and at 45 km still ≈ 2 times higher NO_2 abundances can be measured compared to the standard values at these altitudes during undisturbed conditions. Below altitudes of 40 km, hardly any increase could be observed that was produced directly due to the proton impact. Above 40 and up to about 70 km, the SPE affects NO_x significantly. The NO species are very long lived at these altitudes and the increased values turn back to normal not before a few days after the SPE (see Fig. 5.14). At higher altitudes, starting with about 80 km and higher, the NO_x increase turns back again to normal values very soon. The big peak of NO_x values produced by the SPE only occurs directly during the days of strong highly energetic proton impact. These results are in good agreement with studies using measurements of several other instruments e.g Solar Backscatter Ultraviolet (SBUV/2) instrument and Atmospheric Chemistry Experiment (ACE) [Jackman et al., 2005b]. Very important in this field are the work and publications from Jackman et al. [2005b,

5.3. Results (SPE)

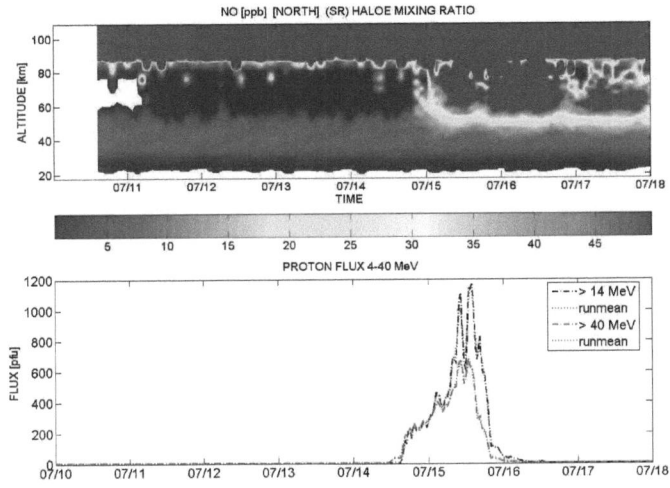

Figure 5.12: *Upper panel: Contour plot of NO during a SPE in July 2000, altitude 12 - 100 km, latitude during SPE ≈ 69° N. Lower panel: Proton flux data provided by GOES. Two energy levels, 9-15 MeV (black dashed line) and 15-40 MeV (red dashed line), blue lines are running mean values.*

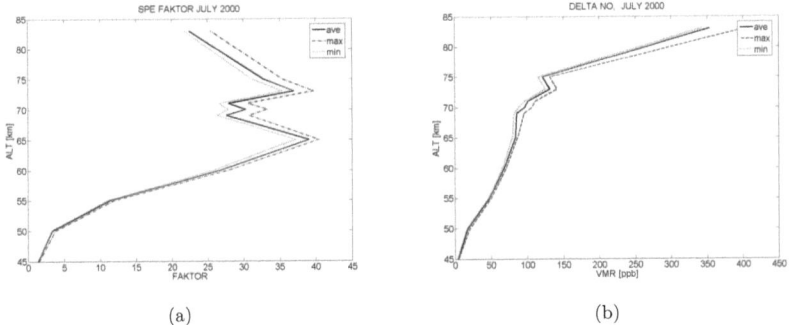

Figure 5.13: *Influence of the SPE from 14th to 17th of July 2000 on the middle atmosphere. a) SPE factor for July 2000. Factor calculated by dividing NO daily average data of disturbed atmosphere profile through the undisturbed profile in July 2000. b) SPE difference NO mixing ratio during July 2000. Standard NO profile subtracted from the disturbed profile during the SPE.*

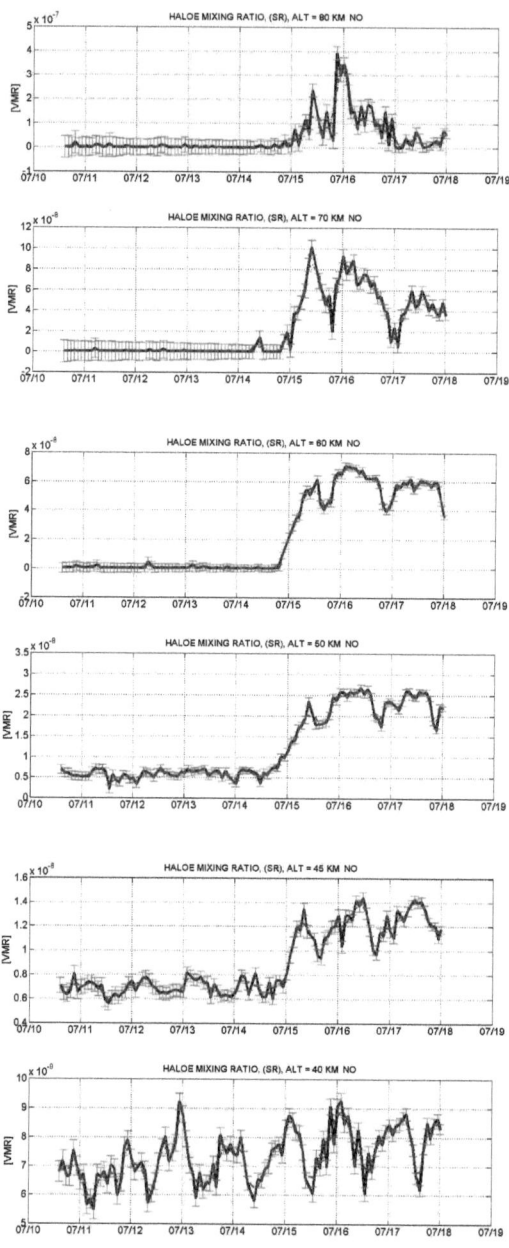

Figure 5.14: *NO [vmr] for some selected altitudes (80, 70, 60, 50, 45 and 40 km), during 10th and 19th July 2000, at geomagnetic latitudes > 60° North.*

5.3. Results (SPE)

2001]. Also in very good agreement are comparisons of the HALOE data with the hybrid ion model developed at the University Bremen. In case of NO the model was able to reproduce nearly the same NO enhancements at the same times and altitudes, see chapter 5.3.7.

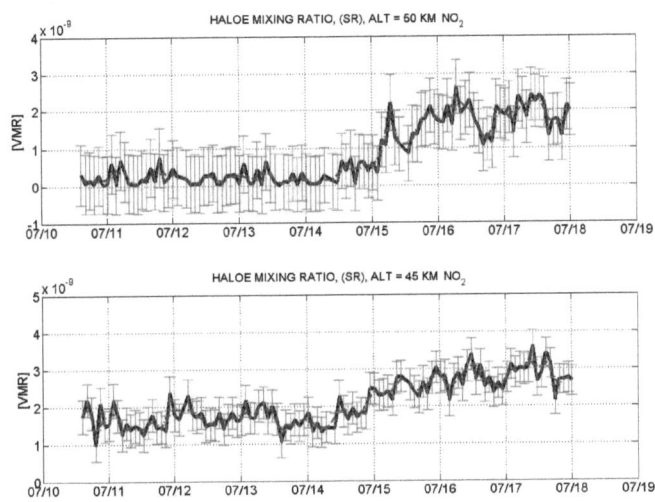

Figure 5.15: NO_2 vmr for some selected altitudes (50 and 45 km), during 10th and 19th July 2000, at geomagnetic latitudes $> 60°$ North.

5.3.4 Bastille Event - O_3

The interaction of atmospheric compounds and the highly energetic protons resulting in production of HO_x (H, OH, HO_2) and NO_y (NO, NO_2, NO_3, N_2O_5, HNO_3, HO_2NO_2, HONO, $ClONO_2$, $ClNO_2$, $BrONO_2$) either directly or through photochemical sequences was discussed in section 5.2 and papers like Jackman et al. [2005a] and many others. The observed ozone depletion monitored by HALOE during the Bastille event is shown in Fig. 5.16. Here, a smooth decrease of ozone between 60 and 45 km can be observed easily from the 15th to 16th of July. The results of several selected altitudes in Fig. 5.17 show massive O_3 depletion during these days e.g at 60 km within a decrease from 12×10^{-7} to 4×10^{-7}. At these altitudes especially the HO_x catalytic cycle is very effective and also great amounts of HO_x are produced by the SPE leading to the observed ozone depletion. The maximum decrease of ozone can be found between 60 and 50 km and reaches values of about 50% strongly decreasing with altitude to about 12% at about

40 km (see Tab. 5.12). At altitudes of about 40 km, the NO_x cycle becomes dominant, thus the ozone depletion at these altitudes is caused mainly by the abundance of NO_x produced due to the SPE. Directly after the event the recovery process of ozone sets in very fast. Nearly undisturbed vmr can be measured a few days after the event again. These results confirm the studies of ozone depletion from Jackman et al. [2005b] investigating the SPE in October 2003.

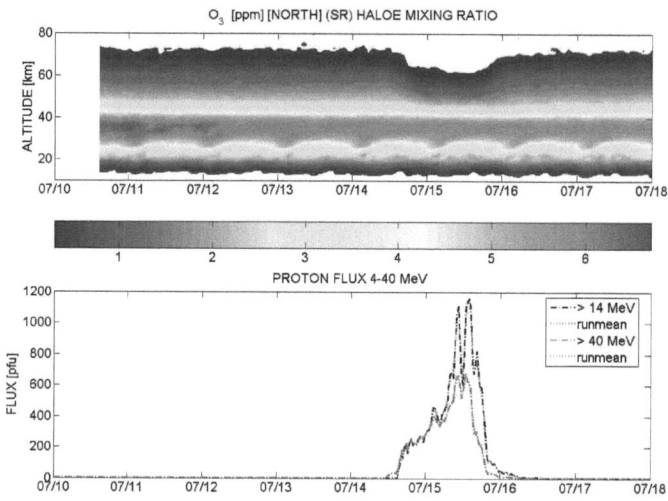

Figure 5.16: *Upper panel: Contour plot of ozone during the SPE in July 2000, altitude 10 - 80 km, latitude during SPE $\approx 69°N$. Lower panel: Proton flux data provided by GOES. Two energy levels, 9-15 MeV (black dashed line) and 15-40 MeV (red dashed line) are shown, blue lines are running mean values.*

5.3.5 Bastille Event - HCl

As mentioned in section 5.3.2, HCl forms a very stable and constant reservoir for chlorine which does not change significantly between 30 and 60 km whether by season or by latitude. Standard vmr values between these altitudes are between approximately 2 $\times$ 10^{-9} and 4 $\times$ 10^{-9}, see also Fig. 5.10. But during the SPE in July 2000 (see Fig. 5.18) during the 15th and 16th of July a short but significant decrease of HCl can be observed of up to ≈ 0.8 ppb.

Investigation of several altitudes showed a decrease in HCl between altitudes from 60 to ≈ 40 km (see Fig. 5.19). Comparisons with the hybrid ion model and HALOE daily averaged data showed very good agreement and results can be found in section

5.3. Results (SPE)

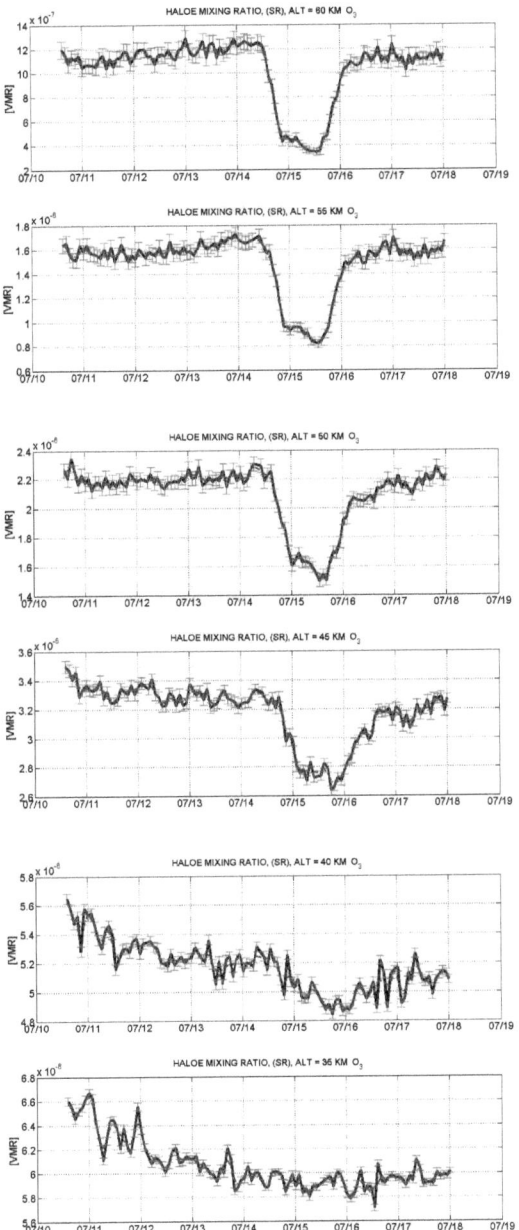

Figure 5.17: O_3 vmr for some selected altitudes (60, 55, 50, 45, 40 and 35 km), from 10th to 19th of July 2000, at geomagnetic latitudes $> 60°$ North.

ALT [km]	SPE [ppm]	NORMAL [ppm]	O$_3$ DECREASE [%]
65	0.2	0.7	~ 71
60	0.3	1.2	~ 68
55	0.8	1.6	~ 50
50	1.5	2.3	~ 35
48	1.9	2.7	~ 30
45	2.7	3.5	~ 23
40	4.9	5.6	~ 12

Table 5.12: *Ozone depletion for several altitudes. Column 'SPE' shows minimum O$_3$ values reached during SPE in July 2000. Column 'NORMAL' shows values before event, column 'O$_3$ DECREASE' gives the Ozone depletion in %.*

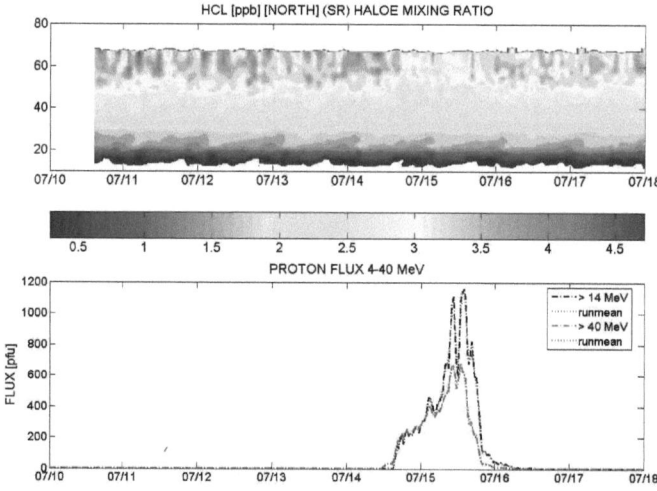

Figure 5.18: *Upper panel: Contour plot of HCl during the SPE in July 2000, altitude 10 - 80 km, latitude during SPE $\approx$ 69°N. Lower panel: Proton flux data provided by GOES. Two energy levels, 9-15 MeV (black dashed line) and 15-40 MeV (red dashed line), blue lines are running mean values.*

5.3. Results (SPE)

6.1.3. Investigations of López-Puertas et al. [2005] and von Clarmann et al. [2005] during the SPE in October 2003 found evidence for chlorine activation from the HCl reservoir by using ClO data measured by MIPAS at latitudes between 70° and 90° North. Their study showed no significant amounts of ClO before the SPE, followed by a sudden increase of ClO during the SPE with maximum values of 0.3 ppb at 42 km. Since no other rapid and efficient source for chlorine was evident, the measured ClO was assumed to be from HCl decomposition released by the reactions in Eq. 5.50 and Eq. 5.51. Taking into account the results of HALOE data, the ion model and the MIPAS measurements, a direct chlorine activation either due to ion cluster chemistry or by increased odd hydrogen abundances due to the SPE seems to be very likely.

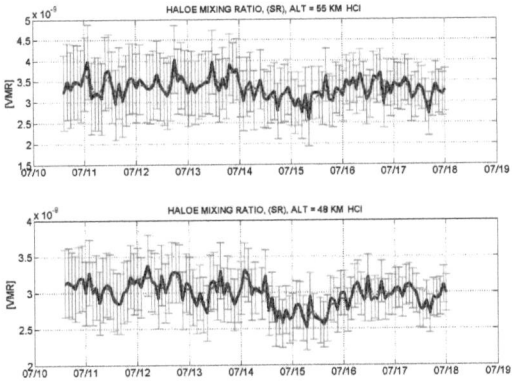

Figure 5.19: HCl [vmr] for some selected altitudes (50 and 48 km) from 10. to 19. July 2000 at geomagnetic latitudes > 60° North.

5.3.6 Temperature Variation

Another question which is not well investigated is wether a change of temperature occurs during SPEs. As the earth's atmospheric temperature profile is strongly connected to wind fields and zonally as well as seasonally altering, investigations of small temperature differences are not conclusive. It is known that ozone is a strong absorber throughout the middle atmosphere, and a sink in ozone leads to less heating rates and thus leads to a local decrease of temperature. On the other hand Joule heating turned out to be a substantial process in the mesosphere leading to an increase of temperature during polar substorms affecting the complex electric field distribution [Jackman et al., 2007].

Fig. 5.20 shows the temperature profile for the month July from ≈ 5 to 120 km altitude covering the latitudes from 80° South to 80° North. The temperature profiles are zonally and monthly averaged and derived from the COSPAR International Reference Atmosphere (CIRA). Studies of Reid et al. [1991] were using model calculations to determine effects of the very intense solar activity period from August to December 1989. During these months extremely high proton fluxes were detected by the NOAA/GOES 7 spacecraft of course leading to enhanced NO_y production and further to ozone depletion especially in stratosphere between 30 - 40 km at polar regions. Their model could reproduce the effects very well and also simulations of temperature effects were made. Results of this model predict temperature decrease of about 3 to 3.5 K between 35 - 45 km during the months October and November compared to undisturbed conditions in former months. Also later studies done by Krivolutsky et al. [2006] using a 3D general circulation and chemical global transport model predicted an SPE induced effect on the temperature budget of the middle atmosphere in July 2000. Their simulations show an induced cooling effect of about 6 K below 80 km and a warming of more than 15 K above 80 km at high latitudes of the northern hemisphere.

For investigations in this study daily averaged temperature data from HALOE were used from the month July 2000. Further the days 11, 12, 13, July were averaged and used as a reference profile, as the data were measured at latitudes between 62° and 65° north and proton fluxes during this time were negligible. The reference profile was subtracted from the HALOE daily averaged data, results are shown in Fig. 5.21, 5.22 and 5.23. The upper panel of Fig. 5.21 shows the differences formed from the reference profile and the daily averaged data. As the SPE effects on NO_x and O_3 could be seen from the 15th of July, variations in temperature can only occur between 15th and 18th of July. Significant is the heating at altitudes between 100 and 140 km, a temperature increase of maximum values up to +185° due to the SPE can be seen. The temperature is slowing down very fast to normal conditions after the maximum of particle event has been reached. To point out that the temperature variations are significant, in Fig. 5.21 the standard deviations of the reference profile is plotted.

Due to the decreasing ozone mixing ratios a lack of the main absorber arises under disturbed conditions, thus a cooling effect is expected at regions where peaks of ozone are formed. As we know, looking at volume mixing ratios, the ozone profile consists of two peaks. The first, also commonly called the ozone layer, is between 30 − 40 km and the second peak at altitudes between 80 − 90 km. Fig. 5.22 upper panel shows only the negative differences, i.e. where cooling occurs. A decrease in temperature of 4 - 6 ° occurs around the 16th of July altitudes between 80 - 100 km, and lasting for about 2 days. Taking into account the standard deviation of the reference profile, errors of ± 1 - 2° do not weaken the significance of the temperature impact. Other temperature decreases at lower altitudes in Fig. 5.22 upper panel could not be found or have values of ≈ 1 - 2°, and thus are probably not significant. In Fig. 5.22 lower panel, the same HALOE data were subtracted from the CIRA model for July, averaging the monthly profiles of the latitudes from 60° to 70°. Comparing these two plots, the same temperature decrease as mentioned above can be seen at the same time and same altitude. In contrast to the values using the HALOE averaged profile, the differences are reaching values up to -3°. Another cooling within the CIRA profile can be seen at the same time below 80 km, and is maybe also caused by the SPE. Temperature variations in the stratosphere seem to be not caused by the particle impact, and might be the result

5.3. Results (SPE)

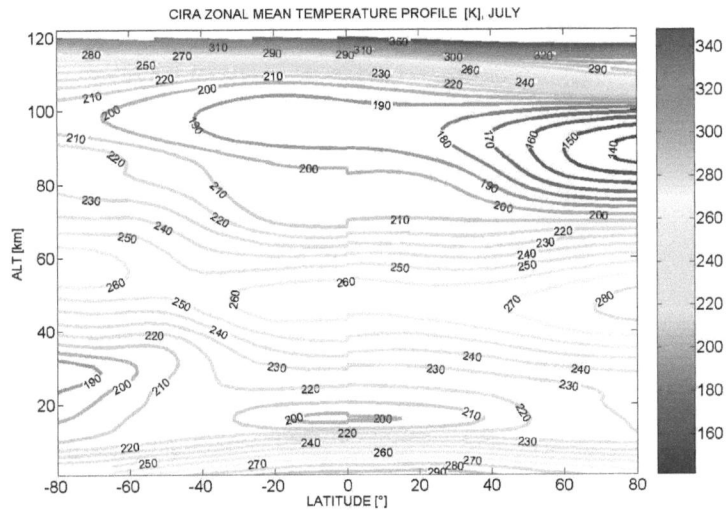

Figure 5.20: *Zonally and monthly averaged temperature field for July in K. Data obtained from the COSPAR International Reference Atmosphere (CIRA).*

of 'natural' variability. There is no clear evidence in Fig. 5.22 that shows significant cooling processes in the stratosphere.
In Fig. 5.23 positive temperature changes are shown, calculated like explained for Fig. 5.22. Here, beside the massive temperature increase in the thermosphere during the SPE, a temperature increase between 3° and 6° K (upper panel) and between 6° and 11° K (lower panel) on the 14th of July at about 80 to 90 km can be observed. This increase is in a good agrement with studies of Jackman et al. [2007] who computed a zonal average adiabatic heating increase in the upper polar southern mesosphere with a maximum of + 2.3° K/day during the Halloween SPE in October 2003.
After looking carefully at previous studies of the effects of SPEs on temperature changes in the middle atmosphere, which are very difficult to observe, the decrease of -2° K (see Fig. 5.22, upper panel) at an altitude of about 60 km seems to be the most realistic effect in terms of cooling processes due to ozone depletion. As an ozone depletion of more than 50% has been observed at this altitudes, a cooling effect at this altitude is very likely.
Problems in terms of investigations of temperature changes from the SPE July 2000 might be caused by the appearance of polar mesospheric clouds (PMCs) at these latitudes. Due to the PMCs the measurements were contaminated and thus might have an effect on the measurements of the used HALOE V19 3AT data set. Perhaps the real

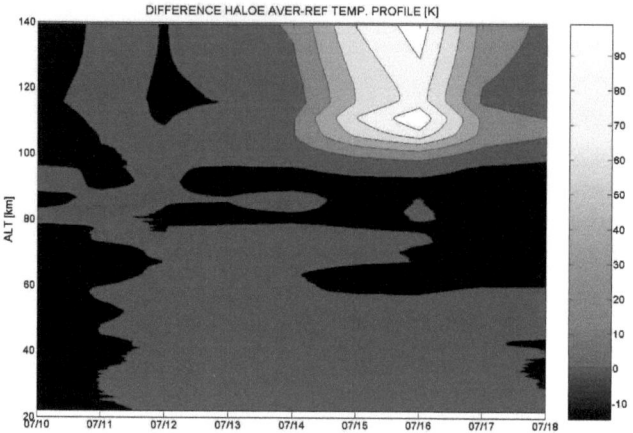

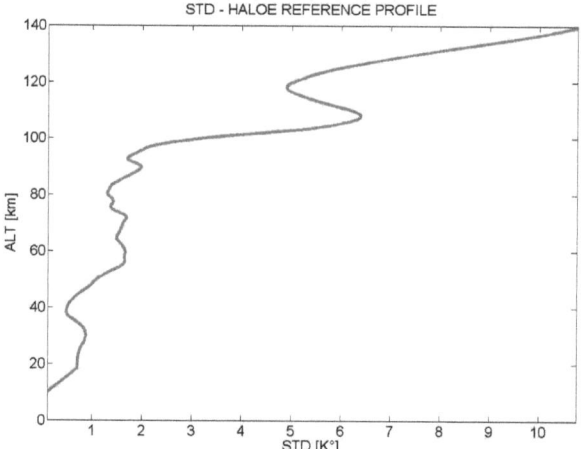

Figure 5.21: *Upper Panel: Positive differences of zonally and daily averaged temperature profile for July 2000 in K from HALOE. The reference profile calculated from 4 days during undisturbed conditions. Lower Panel: Standard deviation of the reference profile.*

temperature effects are disguised by the PMC contamination during the SPE (private communication Dr. C. von Savigny).

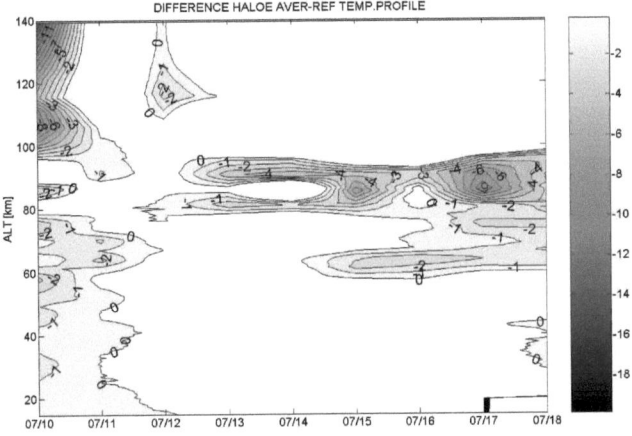

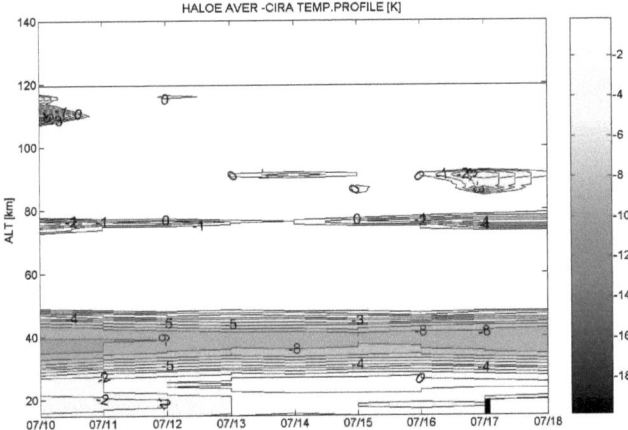

Figure 5.22: *Upper Panel: Negative differences of zonally and daily averaged temperature profile for July 2000 in K from HALOE. The reference profile calculated from 4 days during undisturbed conditions. Lower Panel: Negative differences of zonally and daily averaged temperature profile for July 2000 in K from HALOE with the CIRA standard atmosphere temperature profile.*

80 5. Solar Proton Events (SPE)

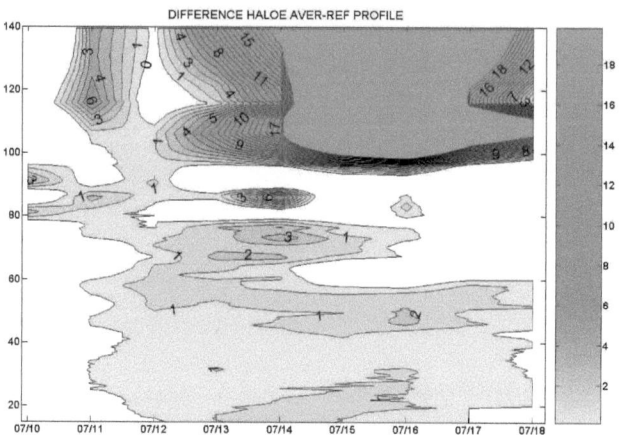

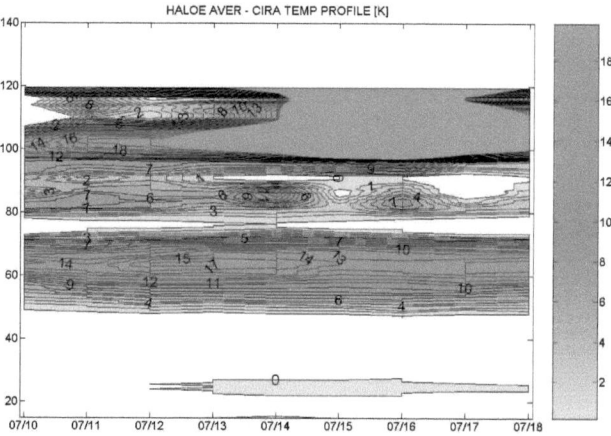

Figure 5.23: *Upper Panel: Positive differences of zonally and daily averaged temperature profile for July 2000 in K from HALOE. The reference profile calculated from 4 days during undisturbed conditions. Lower Panel: Positive differences of zonally and daily averaged temperature profile for July 2000 in K from HALOE with the CIRA standard atmosphere temperature profile.*

5.3.7 SPE - Summary

Large Solar proton events occur mostly during phases of high solar activity and lead to an output of highly accelerated particles as well as emissions of nearly the full electromagnetic spectrum from γ - rays down to radio bursts.

Investigation of the long term data set provided from the HALOE instrument, covering the years 1991 to 2005, showed a clear and strong impact of highly energetic protons especially in terms of the NO_x compounds at geomagnetic latitudes $> 60°$. A significant increase of NO_x down to altitudes of 45 km could be observed in the HALOE data e.g during the Bastille Event in July 2000 and the Halloween Event in October/November 2003. The increase of NO, of up to the factor of 35 at altitudes of 80 km, lasted for the period of the event but returns to normal quite quickly after event at altitudes > 80 km. Below 80 km, NO_x is more stable and the recovery process takes a few days longer.

The increased NO_x further led to a significant decrease of stratospheric ozone at altitudes between roughly 60 and 40 km, during the period of the SPE. HALOE O_3 measurements showed a clear decrease of up to 50 % at altitudes of about 50 km, see Tab. 5.12.

HALOE data also suggest an decrease of HCl during the SPE in July 2000. An decrease of the inactive chlorine reservoir of about 0.5 ppb could be observed and is confirmed by model results (see section 5.3.7). Also studies by von Clarmann et al. [2005] give evidence for simultaneous increase of both ClO and HOCl which suggest a HCl destruction either through OH or directly via ion cluster chemistry. Finally the impact of SPEs on the temperature balance of the middle upper atmosphere was investigated and showed also massive effects e.g. in the lower thermosphere. Here, a significant heating at altitudes between 100 and 140 km of up to +185° could be observed during the Bastille Event. The effect of stratospheric cooling caused by SPEs could not be observed, but a significant decrease of temperature between altitudes of 80 to 100 km can be seen by comparison with a reference profile as well as with comparisons of the CIRA standard atmosphere model.

6. Model vs. Observation

Due to the complexity of the atmosphere, simulations are a well known source for improving the understanding of chemical as well as transport processes throughout the atmosphere. Thus this chapter introduces the hybrid ion model developed at the University Bremen, and shows comparison of species like NO_x, O_3 and HCl during the SPEs and results calculated by the model.

6.1 The Hybrid Ion Model (HIM)

The simulation of solar particle events and their effects on the upper and middle atmosphere as well as the investigation of the resulting ozone loss on a longer time scale was the aim of a study done by Winkler et al. [2008b]. For this purpose a combination of two different kinds of model codes was created. This atmospheric chemistry model consists of the chemistry code of the SLIMCAT model [Chipperfield, 1999] as the photochemical module and the 2D meteorological module THIN AIR [Kinnersley, 1996] which calculates daily mean temperature and pressure fields, and the transport of the chemical species on isentropic levels (see Fig. 6.3) which was further developed by a parametrization from the ionization rates of the main species NO_x and HO_x during the SPE. The vertical resolution is about 3 km up to approx. 100 km height with a latitudinal resolution of 9.47°. The chemistry code simulates gas phase, heterogeneous and photolysis reactions of 57 species. For realistic simulations also anthropogenic emissions of Chloro-fluoro-carbons (CFCs) and green house gases were set to represent the current industrial atmosphere. The transport calculations were triggered by 3 hour steps and the short lived species are treated not independently but in families (NO_x, HO_x, O_x, ClO_x, BrO_x), as the family approach is appropriate in the stratosphere and lower mesosphere. The non - family concept for chemical species is applied for regions above the lower mesosphere (above ≈ 51 km) where the species listed in Tab. 6.13 are treated independently. This hybrid concept led to a much better agrement with measurements data at higher altitudes in contrast to models which are only using the family approach. More detailed model description can be found in Winkler [2007] and Winkler et al. [2008b].

Although the model was designed for investigations of long term effects of SPEs as well as coupling processes with stratospheric ozone compounds over the last centuries and the influence of different geomagnetic field variations, it turned out to be an appropriate tool for computing short term effects and the coupling with species like e.g. chlorine, too. Therefore comparisons with NO_x and O_3 as well as hydrogen chloride (HCl) (pro-

Table 6.13: *The chemical species in the non-family SLIMCAT model.*

Short lived species	O_3, $O(^3P)$, $O(^1D)$, N, NO, NO_2, Cl, ClO, Cl_2O_2, Br, BrO, HCl, HOCl, OClO, $ClONO_2$, HBr, HOBr, BrCl, CH_2O, NO_3, N_2O_5, HNO_3, HNO_4, HO_2NO_2, H_2O_2, $BrONO_2$, H, OH, HO_2, CH_3O_2, CH_3O, HCO, CH_3OOH, H_2O
Long lived species	CH_4, N_2O, CO, $CFCl_3$, CF_2Cl_2, CHF_2Cl, $C_2F_3CL_3$, CH_3Cl, CH_3CCl_3, CCl_4, CH_3Br, $CBrClF_2$, $CBrF_3$, $COFCl$, HF, COF_2
Fixed	N_2, O_2, H_2

vided by HALOE) during large SPE e.g. during the July 2000 event were computed and compared in this study. The effect of precipitating energetic protons and α-particles is prescribed by NO_x and HO_x production rates depending on the number of ion pairs produced. The ion pair production rates are provided by Monte Carlo simulations of ionizing and dissociative interactions of the protons and α-particles with air molecules. A detailed description of the method is given by Schröter et al. [2006], the ionization rates for some selected altitudes for the SPE in July 2000 and October 2003 are shown in Fig. 6.1 and Fig. 6.2. Proton fluxes were obtained from the highly energetic particle detectors onboard the NOOA/GOES satellites and the Interplanetary Monitoring Platform (IMP) and were included into the model as energy source. Due to this new hybrid model, which treats the chemical species as families up to altitudes of the stratopause (≤ 50 km) and at altitudes > 50 km separately, developed and described by Winkler [2007], a proper simulation of the middle atmosphere can be executed.

6.1.1 Results for NO and O_3

To compare model results and NO measurements, daily average NO data were calculated during the Bastille event in July 2000. Due to the measurement technique of HALOE, data obtained in sunrise and sunset mode and were measured at solar zenith angle of 90°. In order to compare the simulation results with the occultation measurements from HALOE, model outputs at solar zenith angles near 90° have been used. The longitudinal dependency of the satellites data is accounted for by 24 model runs performed with hourly shifted ion production rates. From the 24 simulation results a weighted sum has been calculated. By using this approach the errors should be clearly small and also aspects of transport can be neglected by comparing data during such short timescales. Comparison of model runs and HALOE data have been made for all three versions as mentioned before. As the hybrid ion model showed best results, this study will only focus on the results obtained by the hybrid ion model (HIM).

The tremendous effects regarding the production of NO caused by SPE in July 2000 were discussed in chapter 4.4 and are once again shown in Fig. 6.4(a) (upper panels). The values are daily averaged NO [vmr] values observed in sunrise mode and measured

6.1. The Hybrid Ion Model (HIM)

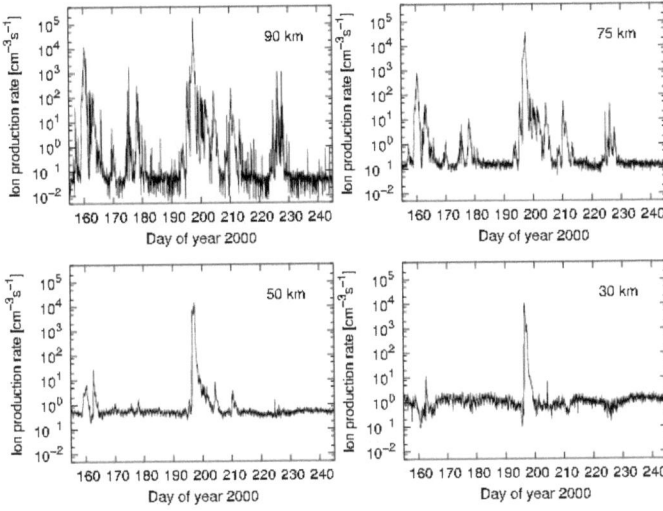

Figure 6.1: *Hourly averaged ion pair production rates at 30, 50, 75 and 90 km altitude during June 1, and September 1, 2000, courtesy of Friedhelm Steinhilber. The Bastille SPE occurred during day 190 and 200. The day 196 corresponds to July 14th. These data have been calculated by the Monte Carlo method using proton and He_2^+ fluxes from GOES. Plot adopted from Winkler [2007].*

inside the polar cusps at geomagnetic latitudes > 60° north. Clearly the strong increase of NO down to ≈ 50 km can be seen starting July 14, 2000 and the fast recovery process with the end of the SPE about 3 days later. The plots in the lower panel of Fig. 6.4(c) show the hybrid ion simulation for the same event calculated for latitudes of ≈ 65.8°. At a first glance the model is in very good agrement with the real data, except at the uppermost boundary, between 80 and 100 km, where the nitric oxide is underestimated already days before the July 14, 2000 due to currently poor estimation of thermospheric NO sources. The low NO thermospheric injection estimations in this model occur because they were taken from observations during the year 1999 during rather quite mean conditions, thus leading to a smaller NO deposition in the lower thermosphere. During undisturbed conditions e.g. the days of the 12th and 13th July, all model runs are tending to exceed the measured NO values in the range 4-6 ppb in the mesosphere, at altitudes between 40 and 80 km. Although the vmr of NO during undisturbed conditions seems to be overestimated in the modeled NO differences, here labeled as Δ NO Fig. 6.4(b) for HALOE NO and Fig. 6.4(d) for the hybrid ion model, model and HALOE data are in good agreement, especially in mesospheric regions (see also Winkler [2007]). At lower altitudes e.g in the upper stratosphere once again an overestimation of NO seems to occur. The differences were calculated by creating an undisturbed averaged profile, calculated from the days of the 12th to 14th of July, which

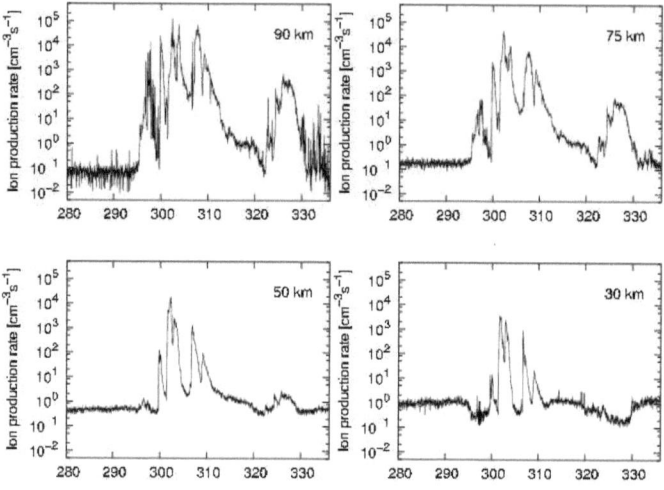

Figure 6.2: *Hourly averaged ion pair production rates at 30, 50, 75 and 90 km altitude during October 6, and December 5, 2003, courtesy of Friedhelm Steinhilber. The Halloween SPE occurred during day 303 and 305 followed by the second phase between day 308 and 310. These data have been calculated by the Monte Carlo method using proton and He_2^+ fluxes from GOES. Plot adopted from Winkler [2007].*

is deducted from the whole data set.

The chemical coupling processes due to the catalytic cycles and the resulting ozone depletion in the stratosphere and mesosphere (Eq. 6.53), have been described by Lary [1997] where X stands for the free catalyst e.g H, OH, NO, Cl or Br. These ozone depleting cycles are also included in the hybrid ion model.

$$X + O_3 \rightarrow XO + O_2$$

$$\underline{XO + O \rightarrow X + O_2} \qquad (6.53)$$

$$\textbf{Net: } O + O_3 \rightarrow 2O_2$$

Fig. 6.5(a) shows daily ozone vmr averaged values obtained from HALOE during the SPE in July 2000, measured in sun rise mode. Clearly a sink in the ozone can be seen during the SPE from the July 14 to 16, 2000 at altitudes between 80 and 50 km but are recovering very fast again. The HIM model Fig. 6.5(c) does not match very well with observations at altitudes over 80 km, but is in good agreement in the

6.1. The Hybrid Ion Model (HIM)

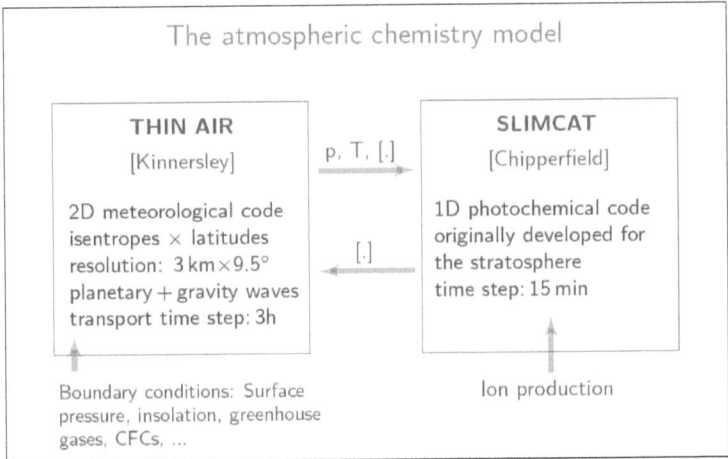

Figure 6.3: *Sketch of the atmospheric chemistry model, combining the THIN AIR module and the SLIMCAT module. (Sketch adopted from Holger Winkler's defense presentation).*

middle and upper mesosphere between 70 km and 80 km. Values in the stratosphere cannot be reconstructed very well and exceed the measured values of ozone even during undisturbed conditions. By comparing the total differences of HALOE and the HIM model Fig. 6.5(b)- 6.5(d) a better agreement could be achieved and proofs to be a good proxy for investigations of short effects like e.g. SPE.

6.1.2 University Bremen Ion Chemistry Model (UBIC)

The effects of SPEs and their influence on NO_x increase and following O_3 depletion was shown in the section 6.1.1 by satellite data as well as by simulations carried out with the HIM model. Solomon and Crutzen [1981] pointed out a possible interaction of the increased NO_x and especially HO_x concentrations with chlorine species in the lower mesosphere and stratosphere. The impact due to the SPE leads to a formation of chlorine nitrate at the expense of reactive radicals by the reaction Eq. 6.54:

$$ClO + NO_2 + M \longrightarrow ClONO_2 + M \tag{6.54}$$

The effect of this reaction was found to be significant in the lower and middle stratosphere but turned out to be less important looking at higher altitudes [López-Puertas et al., 2005]. However, studies by von Clarmann et al. [2005] regarding the SPE in October/November 2003 of short enhancements of ClO and HOCl concentrations in

NO Hybrid Ion Model vs. HALOE

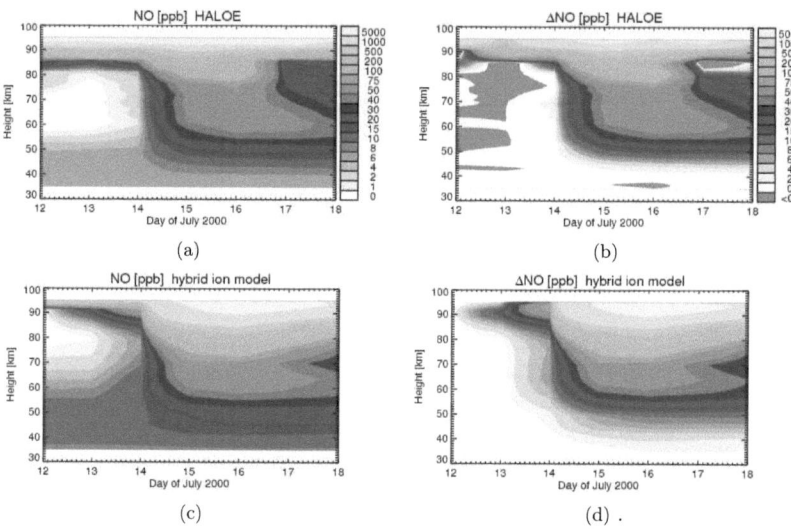

Figure 6.4: *NO increase due to the SPE in July 2000. Comparisons of hybrid ion model and HALOE data.* **a)** *NO daily average values during SPE in July 2000, sunrise mode, latitude during SPE $\approx 69°$, NO mixing ratio in ppb.* **b)** *NO differences from HALOE daily averaged data in July 2000, sunrise mode, latitude during SPE $\approx 69°$, NO mixing ratio in ppb.* **c)** *NO model simulation of SPE July 2000. NO mixing ratio in ppb.* **d)** *NO difference model simulation during the July 2000 event. Delta NO mixing ratio in ppb.*

the polar regions showed another particularly important process by transformation of hydrogen chloride into reactive chlorine species by Eq. 6.55 :

$$HCl + OH \longrightarrow Cl + H_2O \tag{6.55}$$

It was found, using measurements of the Michelson Interferometer for Passive Atmospheric Sounding (MIPAS) onboard the Environmental Satellite (ENVISAT), that this reaction seems to be most effective at altitudes between 40 km and 45 km, due to short term enhancements of ClO and HOCl. It was also mentioned that possibly ion chemistry reactions could lead to a conversion of HCl to reactive chlorine species. Several ions e.g. CO_3^-, O_2^-, O^-, NO_2^- and NO_3^- are reacting with HCl to produce Cl^-. Further the Cl^- are forming $Cl^-(X)$ clusters where X= (HCl, H_2O, CO_2) [Fritzenwallner and Kopp, 1998] and [Winkler et al., 2008a]. For purpose of HCl comparisons of HALOE HCl data and the hybrid ion model, mentioned in section. 6.1, the model has been improved. One of the main differences is the replacement of the equilibrium approach of the original SLIMCAT code by an independent treatment of the chemical species e.g.

6.1. The Hybrid Ion Model (HIM)

O_3 Hybrid Ion Model vs. HALOE

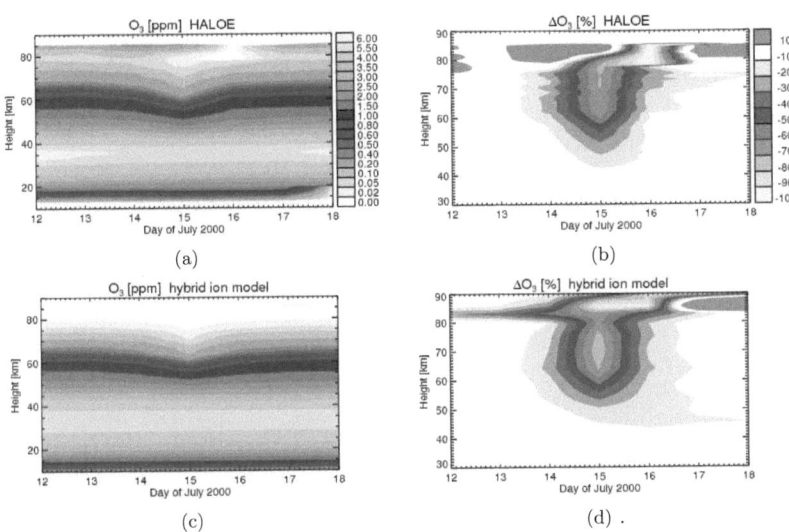

Figure 6.5: *Ozone depletion during the SPE in July 2000. Comparisons of hybrid ion model and HALOE data. a) O_3 daily average values during SPE in July 2000, sunrise mode, latitude during SPE ≈ 69°, O_3 mixing ratio in ppm. b) O_3 difference from HALOE daily averaged data in July 2000, sunrise mode, latitude during SPE ≈ 69° in %. c) O_3 model simulation of SPE July 2000. O_3 mixing ratio in ppm. d) Delta O_3 model simulating the July 2000 event. O_3 difference mixing ratio in %.*

the families NO_x, HO_x, BrO_x, ClO_x and O_x Winkler et al. [2008a]. UBIC simulates the time evolution of 138 species with about 550 reactions (plus recombination) and also accounts for photo ionization of NO by Lymann-α radiation, photo dissociation of charged species and photo detachment of electrons (see Tab. 6.14).

6.1.3 Results HCl

To compare once again model and satellite data, outputs of the UBIC model in sun rise mode were used, as data obtained from HALOE during the SPE were also measured in sun rise mode. Thus, 24 model runs were performed with hourly shifted ionization rates. HALOE HCl data as well as the model are representing volume mixing ratio values inside the polar cusps (geomagnetic latitudes ≥ 60°) during the SPE in July 2000. Fig. 6.6 shows the observed and simulated HCl differences during the 11th and 18th of July in the year 2000. Here the solid line shows the UBIC simulation results, the dashed line with big squares represents the HALOE daily averaged data and the dashed dotted line shows results of a model run including ion chemistry using the

Table 6.14: *UBIC species*

Cations	N^+, N_2^+, NO^+, NO_2^+, $O^+(^4S)$, $O^+(^2D)$, $O^+(^2P)$, O_2^+, $O_2^+(a^4)$, O_4^+, O_5^+, H^+, CO^+, CO_2^+, HCO^+, H_2O^+, $O_2^+(H_2O)$, $H^+(H_2O)_{n=1...7}$, $H^+(H_2O)(OH)$, $H^+(H_2O)(CO_2)$, $H^+(H_2O)_2(CO_2)$, $H^+(H_2O)(N_2)$, $H^+(H_2O)_2(N_2)$, $H^+(CH_3CN)$, $H^+(CH_3CN)(H_2O)_{n=1...6}$, $H^+(CH_3CN)_2$, $H^+(CH_3CN)_2(H_2O)_{n=1...4}$, $H^+(CH_3CN)_3$, $H^+(CH_3CN)_3(H_2O)_{n=1,2}$, $NO^+(H_2O)$, $NO^+(H_2O)_2$, $NO^+(H_2O)_3$, $NO^+(CO_2)$, $NO^+(N_2)$, $NO^+(H_2O)(CO_2)$, $NO^+(H_2O)_2(CO_2)$, $NO^+(H_2O)(N_2)$, $NO^+(H_2O)_2(N_2)$, $NO_2^+(H_2O)_{n=1,2}$
Anions	e, O^-, O_2^-, O_3^-, O_4^-, OH^-, NO_2^-, NO_3^-, CO_3^-, CO_4^-, CH_3^-, HCO_3^-, $O^-(H_2O)$, $O_2^-(H_2O)_{n=1,2}$, $O_3^-(H_2O)_{n=1,2}$, $OH^-(H_2O)_{n=1,2}$, $NO_2^-(H_2O)_{n=1,2}$, $NO_3^-(H_2O)_{n=1,2}$, $CO_3^-(H_2O)_{n=1,2}$, $NO_3^-(HNO_3)_{n=1...4}$, $NO_3^-(HNO_3)(H_2O)$, $NO_3^-(HNO_3)_2(H_2O)$, $H_2SO_4^-$, $HSO_4^-(H_2SO_4)_{n=1,2}$, $HSO_4^-(H_2SO_4)(H_2O)$, $HSO_4^-(HNO_3)_{n=1,2}$, $HSO_4^-(HNO_3)(H_2O)$, $HSO_4^-(HNO_3)_2(H_2O)$, $HSO_4^-(H_2SO_4)(HNO_3)$, $HSO_4^-(H_2SO_4)(HNO_3)(H_2O)$, Cl^-, Cl_2^-, Cl_3^-, ClO^-, $ClO^-(HCl)$, $ClO^-(H_2O)$, $ClO^-(CO_2)$, $ClO^-(HO_2)$, $NO_3^-(HCl)$
Neutrals	$N(^4S)$, $N(^2D)$, N_2, $O(^3P)$, $O(^1D)$, O_2, O_3, H, H_2, OH, HO_2, NO, NO_2, NO_3, N_2O, H_2O, CH_4, CH_3, CO_2, CO, HCO_3, HNO_3, HNO_2, N_2O_5, H_2SO_4, CH_3CN, Cl, Cl_2, ClO, $ClNO_2$, $ClONO_2$, HCl, $HOCl$

parametrization for the HO_x and NO_x obtained by Rusch et al. [1981] and Porter et al. [1976]. The UBIC model shows better results compared to the model run using only parametrization. At all altitudes large differences between simulation and observed data can be seen. UBIC seems to reproduce HCl differences at altitudes ≥ 45 km well. Further model runs proof, that the increase of species e.g. ClO, Cl and HOCl correspond to the HCl decrease during the SPE in July 2000 as well as at the SPEs in October/November 2003. As those SPEs are comparable because they were of roughly the same strength [Jackman et al., 2005a], it is reasonable to compare these events. The modeled increase of Cl and HOCl after both "big SPEs" turned out to be of the same order. Differences between model and observation could mainly be found because the simulations predicted that the transformation from HCl into active chlorine is a short lived effect, but observations seem to indicate a slower recovery of the HCl species. Finally Fig. 6.7 shows observed and modeled zonally averaged profiles of HCl differences for both model and the HALOE data versus altitude. The profiles are calculated for the July 16, 2000, as the impact of protons reached its maximum. As can be seen the UBIC model shows the best agreement compared to the observed data. The parameterized run (PARAM) does not seem to fit at all.

6.1.4 Model Errors

Of course models can only be a good approximation of reality therefore several error sources are included in model results. The model only accounts for a limited set of processes and does this on a finite two dimensional simulation grid. Model results showed unrealistically large total ozone columns (TOC) for polar region due to the

6.1. The Hybrid Ion Model (HIM)

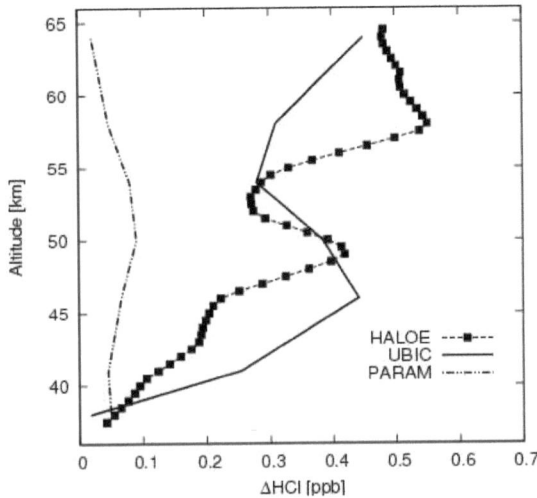

Figure 6.6: *Observed and modeled zonally averaged HCl differences ppb at sun rise on July 16th (with respect to the mean HCl mixing ratio of 11th to 13th of July) as a function of altitude. Shown are model results with UBIC and with parameterized production rates (PARAM) for HO_x and NO_x, respectively.*

chosen transport approach. The chemical scheme of SLIMCAT consists of more than one hundred reactions which are based on reaction coefficients originated from measurements or even experiments containing a big range of uncertainties. The calculated errors amount to between 1 - 4.5 % in TOC. The most critical input parameter for the simulation are the ion pair production rates, which are leading to uncertainties of $\approx$ 10 %. For detailed information on error estimations of the HIM see Winkler [2007] and in terms of the UBIC see Winkler et al. [2008a].

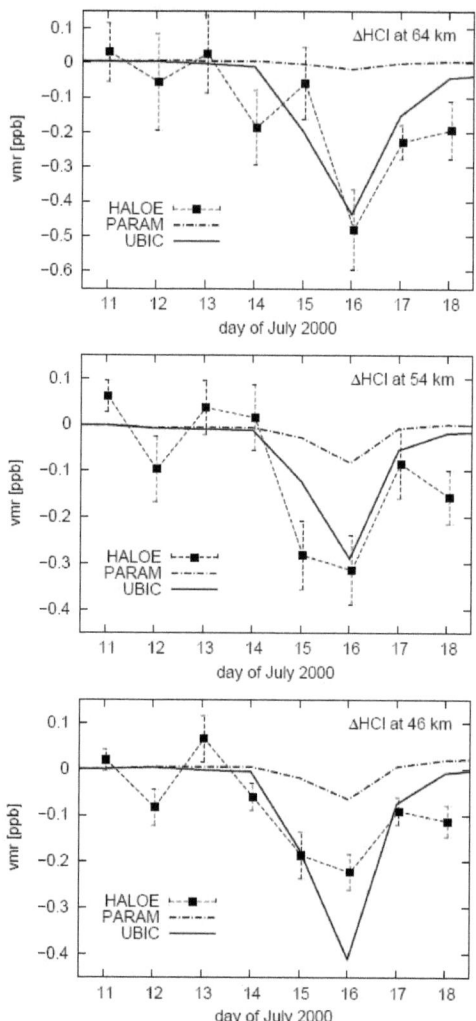

Figure 6.7: *Differences of zonally averaged HCl mixing ratios at three different altitudes (64 km, 54 km and 46 km) during sun rise conditions (differences with respect to the mean HCl values of 11th to 13th of July). Shown are HALOE data, simulation results from the atmospheric model at 66.5° North using UBIC, and using the parameterized production rates for HO_x and NO_x, respectively [Winkler et al., 2008a].*

7. Highly Energetic Electron Precipitation (EEP)

In this chapter three questions are addressed:

- What is the best proxy for EEPs, (GOES, POES or Ap-index)?
- Does the EEP direct effect have a significant influence on the middle atmosphere, and does it affect the atmosphere in the same way as SPEs?
- How large is the so called EEP indirect effect and can it explain the observed polar winter O_3 anomalies?

To quantify the influence of these processes, correlation coefficients were calculated for altitudes from the stratosphere up to the lower thermosphere. All results presented in this chapter have undergone statistical tests (e.g. the Student's t-test) to establish their statistical significance.

7.1 Electron fluxes: GOES, POES and Ap-index

Discussions during a special meeting focussing on the highly energetic particle precipitation (HEPPA meeting in May 2008), led to questions regarding the optimal data source for investigations of highly energetic particle fluxes regarding the highly energetic electron precipitation events (EEP). Thus, in this section the GOES electron flux providing electron fluxes with E > 2 MeV, the POES electron fluxes with energies of 300 keV to 2.5 MeV and the Ap-index are compared. Very important in this context is the comparison of the POES data with the Ap-index, as it is commonly used as proxy for the energetic electron precipitation. In Fig. 7.1 (upper panel), daily averaged values of the Ap-index covering the years 1991 - 2005 are plotted versus POES data of the same time. The formation of three different populations might be seen, two small branches with linear correlation, and one big population with no significant correlation at all. The overall correlation coefficient of R = 0.39 is very small, hence, the daily averaged Ap-index does not seem to be a good proxy for impulsive electron flux changes. On the other hand Fig. 7.1 (lower panel) shows the monthly averaged Ap-index versus monthly averaged POES electron flux. Here, data show a significant linear correlation with an overall correlation coefficient of R = 0.79 during the years 1991 to 2005, which gives clear evidence for a significant dependency of these data. Furthermore, Fig. 7.2 shows

Ap-index and electron flux monthly averaged data plotted versus time. Both of these plots clearly show a very similar shape. Especially the highly active phases in summer 1991, spring 1994 and the very active year 2003 are clearly observed in both data sets. The results show that the daily values of the Ap-index do not seem to be a good proxy for the energetic electron flux, as it might only respond to very highly energetic electrons beyond at least a few MeV. But the Ap-index seems to be a good proxy in terms of monthly averaged values. Thus, the Ap-index turned out to be a sufficient data source for long term observations and is able to point out active as well as inactive phases in periods of months.

The GOES and POES spacecrafts measure at completely different locations concerning the population of charged particles (see sections 4.2 and 4.4). Comparison of the GOES and POES highly energetic electron flux data show no correlation in terms of daily measurements. We obtain a correlation coefficient of R = 0.1 and Fig. 7.3 shows that more electron events are observed in the POES data set than in the GOES data set. Also calculation of the monthly averaged electron values showed a quite poor coefficient of R = 0.46. As both spacecrafts are located in different orbits, different effects in the magnetosphere might be monitored. Another possible reason for these differences was given in private communication with Dr. Craig Rogers (Dept. of Physics, University of Otago). He suggested that the channels of the POES/MEPED detector are insensitive to the range of 0.3 to 2.5 MeV electrons and are dominated by the behavior of those electrons between $\approx$ 0.3 and 0.6 MeV, due to the integral nature of the channel and the power-law distribution of typical radiation belt fluxes.

Another explanation for the higher counts of increased highly energetic particle fluxes by POES is the so called Dst-effect [Kalicinsky, 2008]. This effect occurs due to the fact that geomagnetic storms increase the ring current which leads to a decreased strength of the Earth's magnetic field. If these changes of the magnetic field happen slowly compared to the characteristic time scales of the particle motion, the particles are forced to preserve the adiabatic invariants (see also section 3.4.1). As the decrease of the magnetic field ($\mid B_1 \mid > \mid B_2 \mid$) forces the charged particle to maintain the third adiabatic invariant, the electrons have to move on a higher drift orbit on a higher shell ($L_1 > L_2$) see Fig. 7.4. Also the first and second adiabatic invariants demand a decrease of the electrons energy ($E_1 > E_2$) because of the decrease of the magnetic field due to the increase of the ring current [Onsager et al., 2002]. As the GOES satellite moves on its constant orbit with a constant distance to the Earth, this effect leads to a decrease of electron flux in its measurements. The satellite now observes particles which were moving on a lower orbit with higher energy before the storm. Thus, the measurements of the geostationary spacecraft show a loss of particles in the magnetosphere, although this loss is only insubstantial. As these effects are caused by the increased ring current which further also forces a change of the Dst-index (see Appendix), it was named the Dst-effect. Because of these adiabatic motion effects, data of geostationary charged particle fluxes have to be processed very carefully. This might also be an explanation for the lower electron fluxes in the GOES data set compared to the POES data covering the same time range. Another interesting observation when comparing the GOES and POES highly energetic electron fluxes can be seen in Fig. 7.5. It clearly shows that

7.1. Electron fluxes: GOES, POES and Ap-index

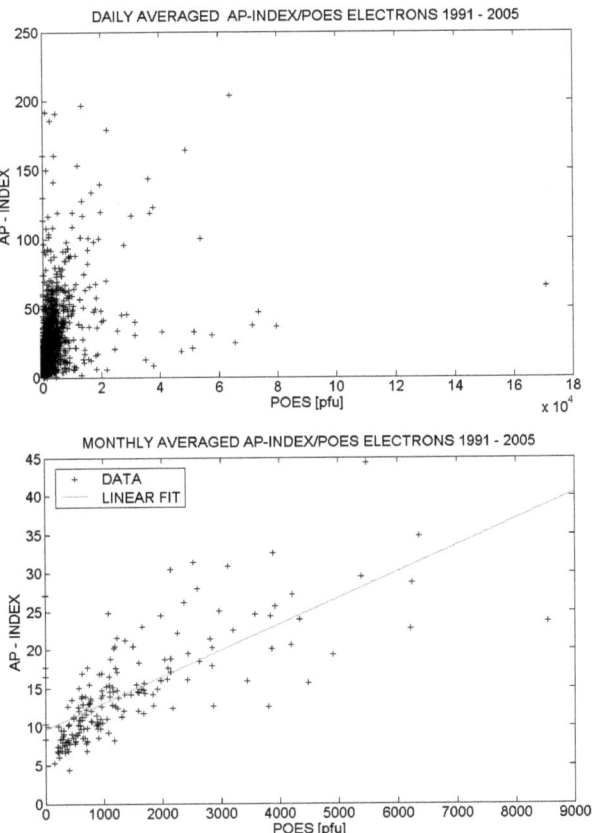

Figure 7.1: *Upper panel:* Daily Ap-index data versus POES electron flux (300 keV - 2.5 MeV) covering the years 1991-2005. Correlation coefficient $R = 0.39$. *Lower panel:* Monthly averaged Ap-index data versus POES electron flux (300 keV - 2.5 MeV) covering the years 1991-2005. Correlation coefficient $R = 0.79$.

especially during the large SPEs in July 2000 and October 2003 as well as during high proton fluxes in the year 2001, no significant increase of electron fluxes are measured by GOES, but they are measured by POES. This effect might also be caused by the Dst-effect mentioned before.
However, since effects of highly energetic particles at polar regions are dependent on those particles entering these regions, POES data and also the Ap-index, in terms of

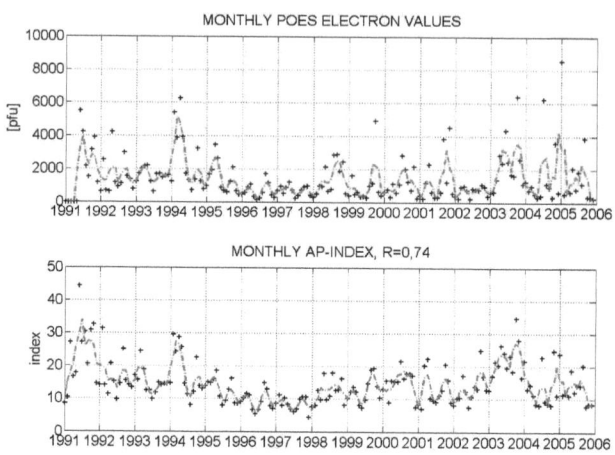

Figure 7.2: **Upper panel:** Monthly averaged POES electron flux (300 keV - 2.5 MeV) covering the years 1991 to 2005 (black crosses). Dotted line shows 3-day running mean. **Lower panel:** Monthly averaged Ap-index (black crosses) covering the years 1991 to 2005, dotted line shows 3-day running mean. Correlation coefficient $R = 0.74$.

long term studies, should be taken into account.

7.2 EEP direct effect

In contrast to highly energetic protons, energetic electrons of the containing the same energy, can penetrate deeper into the atmosphere. This is possible because of their lower mass and size (see also section 3.2) also, they occur much more frequently (see Fig. 7.5). Thus, investigations of EEPs affecting the middle atmosphere, and the question whether the effects are comparable to the effect caused by SPEs are of great interest. In Fig. 7.5 important data of electron fluxes (GOES, POES), proton flux and solar activity are normalized and shown together for a better comparison. Differences between the GOES and POES electron flux data were discussed before in section 7.1. Two different groups of EEPs can be observed. The first group seems to be following large SPEs, thus occur during high solar activity, mostly directly during or a few days after large enhancements of proton fluxes (1991 - 1993 and 2001 - 2002) and are often not observed by GOES electron flux measurements. The other group of large EEPs are monitored at solar minimum, and during inclining and declining solar activity e.g. during the years 1994 - 1996, 1998 - 1999 and 2004 - 2005.

A perfect example for investigations of a direct highly energetic electron precipitation

7.2. EEP direct effect

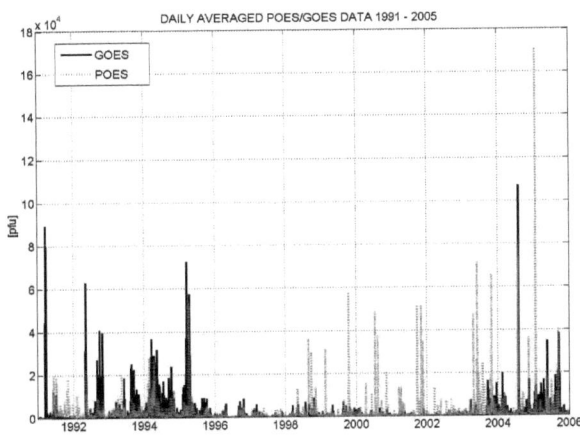

Figure 7.3: *Daily averaged POES electron flux (300 keV - 2.5 MeV) covering the years 1991 to 2005 (dashed line) and daily averaged GOES data (solid line) for the same energy and time range.*

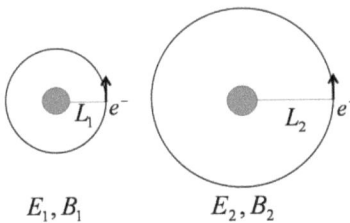

Figure 7.4: *Sketch of the Dst- effect.* **left:** *electron before geomagnetic substurm* **right:** *electron during the substurm .*

event is the EEP that happened in May 1997. Here a big coronal mass ejection (CME) dashed large amounts of charged particles the in direction of the Earth and was detected by the Solar and Heliospheric Observatory (SOHO) on May 12, 1997. Due to measurements of the Solar Wind Experiment onboard of the WIND spacecraft, solar wind velocities of about 500 kms^{-1} and a plasma density increase from 15 to 50 particles cm^{-3} were detected on May 15, 1997 when the EEP reached its maximum [Ogilvie et al., 1995]. During this time the Solar Anomalous and Magnetospheric Particle Ex-

7. Highly Energetic Electron Precipitation (EEP)

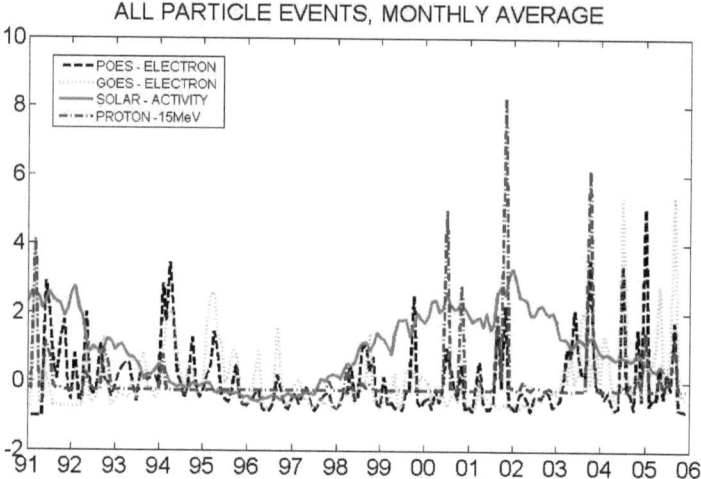

Figure 7.5: *Normalized solar particle flux data and solar activity. POES electron flux measurements covering the years 1991 to 2005 (dashed line). Electron energy levels < 2.5 MeV. Data obtained from Jan Maik Wissing, University Osnabrueck, Department for Numerical Physics. GOES electron flux data covering the years 1991 to 2005. Counts of electron flux with energy levels between 300 KeV and 2.5 MeV (dotted line). GOES proton flux measurements for energy levels 8 - 44 MeV for the years 1991 - 2005 (dashed dotted line). Solar activity derived from Mg II -Index data covering years 1991 to 2005, data obtained from IUP homepage [Weber, 1999] (solid line).*

plorer (SAMPEX) measured an increase of the electron flux by a factor of 4-5 with energy levels of E ≥ 400 keV and also the MEPED instrument onboard POES detected an increase of the electron flux for 300 keV ≤ E ≤ 2.5 MeV by about 4 to 7 times, see Fig. 7.6. The EEP lasted for about 12 days and ended on the 22nd of May reaching moderate electron flux values again. During this EEP no significant increase in proton flux has been measured either by GOES or the POES satellites, thus any effects by highly energetic protons can be excluded. The effect caused by the EEP on NO can be seen very well in HALOE data Fig. 7.6, as during these days the instrument measured at latitudes over 60 degrees north. The relative change of NO throughout the EEP period is calculated relative to the time before the event, creating a reference profile by averaging 50 profiles between 40° and 60° North. Using this method an increase of a factor of over 30 can be seen especially at altitudes of about 90 and 110 km during the days when the EEP reached its maximum. The bulk of energy is deposited at altitudes between 90 and 140 km but does not seem to affect lower regions. Fig. 7.7 shows three

7.2. EEP direct effect

different profiles of NO, one during undisturbed the other two during disturbed conditions, divided by the reference profile. The single profiles measured during the EEP on the 15th and 16th of May clearly show an increase in NO [vmr], compared to the single profile measured during undisturbed conditions, on the May 5, 1997. Also compared to the reference profile a significant increase of NO is observed on the 15th and 16th of May. The NO factor increases up to 25 at altitudes of about 100 km in contrast to the 5th of May and also the vmr values increase dramatically up to values of about 6×10^{-4} at altitudes 110 and 130 km. As the electron flux turns back to normal values on the 21st of May, the NO production also turns back to normal values. Looking at data from days after EEP maximum, the NO budget is nearly turning back to normal conditions within a few days by 21st to 22nd of May. As the produced NO is photochemically removed very fast in the northern hemisphere (because of summertime conditions and sunlight), a downward transport to the stratosphere can not take place and thus this event has no significant influence on ozone depletion. The investigations and modeling of the influence of even lower energetic electron precipitation fluxes due to the EEP May 1997 have also been published by Callis et al. [1998] and show similar results for the northern hemisphere. During the EEP HALOE was observing the northern hemisphere in sun rise mode at latitudes of about 60 and 70 degrees, data in sunset mode were made at latitudes of -50 to -30 degrees and cannot be taken into account for EEP direct effect investigations.

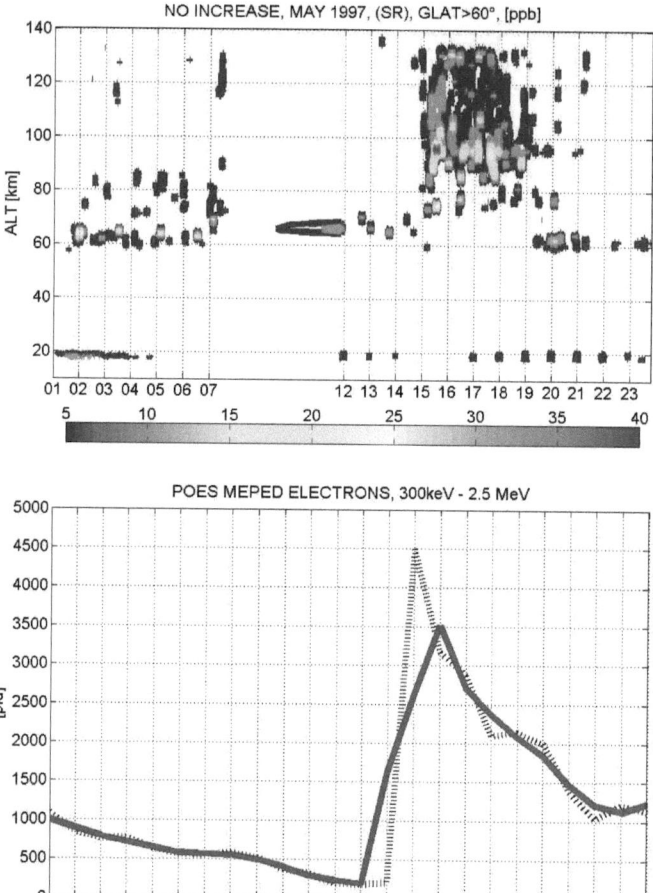

Figure 7.6: *Upper panel:* NO increase derived from HALOE data during May 1997 covering altitudes 20 to 140 km. Maximum increase (calculated by dividing through NO abundances for undisturbed condition) can be seen during maximum of the EEP between 90 and 120 km altitude. *Lower panel:* POES daily averaged electron flux measurements $300\,keV \leq E \leq 2.5\,MeV$ in May 1997 (black dashed line), and 3-day running mean values (blue solid line).

7.2. EEP direct effect

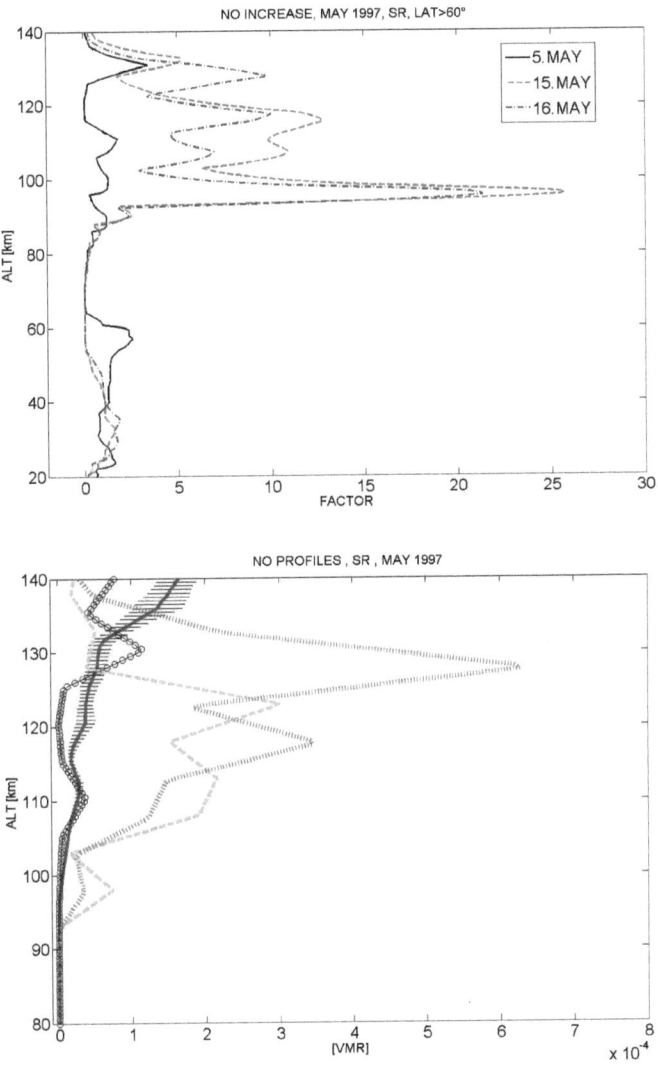

Figure 7.7: *Upper panel: Factor calculated from NO single profiles during the CME on May 12, 1997 derived from HALOE data. Factor calculated by dividing disturbed condition through undisturbed condition. Solid line shows NO on the 5th May during undisturbed condition. The 15th May (dashed line) and 16th May (dotted line) show NO data during the CME.* **Lower panel:** *HALOE NO vmr single profiles. Reference profile averaged from latitudes from 40-60 degrees measurements during undisturbed conditions (blue solid line with error bars showing the standard deviation). Black line with circles shows an undisturbed single profile on the 5th of May, green dashed line shows a profile during the CME on the 15th May and red dotted line also shows an profile during the CME from May 16, 1997. For information of the vertical resolution of HALOE single profiles see Tab. 4.5.*

7.2.1 EEP long time observation

Chapter 5.2 described clearly that highly energetic protons can lead to an increase of NO_x in the middle atmosphere for weeks, and it was pointed out as well, that very often increased proton fluxes are accompanied by an increased flux of highly energetic electrons. Therefore the sources of increased NO_x values are very hard to distinguish, as they are produced either by the electron precipitation or the protons or both. Looking at data of highly energetic electron and proton fluxes of the years 1991 to 2005, very often small SPEs and small EEPs occur in the same month, therefore those time ranges had to be excluded to study the direct effects caused solely by highly energetic electrons. The next challenging step is to select data that has been detected at the right location during such a separated event. That means in the ideal case that only measurements observed at geomagnetic latitudes over 60° north and south can be used, as the electrons are entering through the polar cusps and auroral regions. To find a correlation between these direct effects of highly energetic electrons and enhanced NO production for different altitudes, we were comparing the POES electron flux data with energy range 300 keV $\leq$ E $\leq$ 2.5 MeV and with energy ranges of 100 keV - 300 keV to HALOE NO data made at geomagnetic latitudes $\geq$ 60°. To find out if there is any evidence for linear coupling between the POES electron flux and increased NO in the mesosphere and the lower thermosphere, electron flux data from POES and GOES were plotted against HALOE data observed beyond 60° geomagnetic latitude, as well as at regions of the auroral oval (55°−65°) and at latitudes greater than 60°. Unfortunately, not enough data are available at geomagnetic latitudes greater 60° thus the results showed no clear evidence of a linear correlation with highly energetic electron fluxes (plots are not shown). Thus we focused on the auroral oval regions and data at latitudes greater than 60° geographic latitude.

Fig. 7.9 shows results for NO data plotted versus the electron fluxes of POES for energies of 100 - 300 keV and linear fits obtained by using the linear regression method with implied error estimations (see also Appendix A). Data measured during periods with highly energetic proton fluxes greater than 10 pfu (pfu see Appendix A) were excluded from the data set. Thus, only direct effects of highly energetic electron fluxes should be influencing the NO budget. Data observed in the northern auroral oval (black crosses) and at geographical latitudes greater than 60° north show a linear correlation. The linear fit (solid line) as well as the linear regression fit with error implementation (dashed line) show a linear relation between NO and precipitating energetic electrons between 100 - 300 keV at altitudes between roughly 80 and 110 km altitudes. These relations could not be found by comparison with electron of energy between 300 keV and 2.5 MeV, the result did not show any significant correlation.

Also further calculation of the correlation coefficients of NO and electron flux 100 - 300 keV for the South and North hemisphere in sun rise as well as in sun set mode show a significant correlation in a comprehensive way, see Fig. 7.10. The best results of correlation coefficients greater than 0.5 are observed at latitudes of the auroral oval at both hemispheres. Nearly all results in Fig. 7.10 show a similar shape of correlation with electron fluxes 100 - 300 keV at altitudes between 80 and 110 km, which gives evidence for the EEP direct effect. All studies showed that the direct effect seems to

7.2. EEP direct effect

NO vs. electron flux (SR)

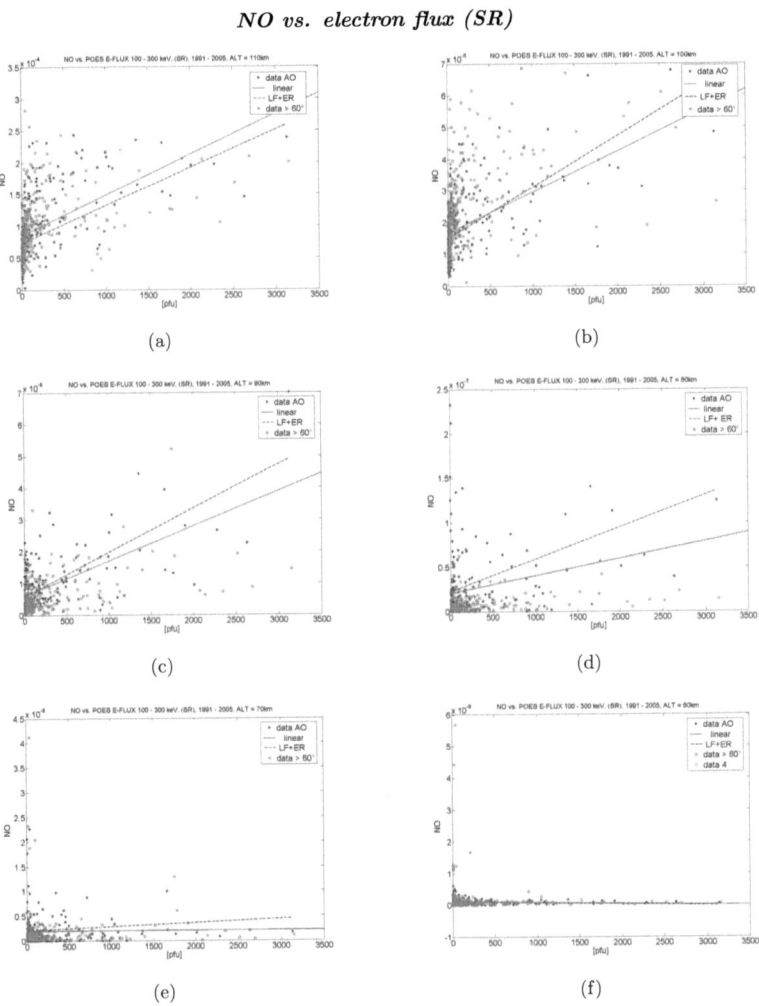

Figure 7.9: *HALOE Daily averaged NO vmr versus POES electron fluxes for the 100 keV - 300 keV energy range from 1991 to 2005 for several altitudes from 110 to 60 km. Only NO data are taken into account during times ranges of very low proton fluxes. Black crosses show data measured in the auroral oval (latitude 55° - 65°), red circles show data at latitudes greater than 60°. Solid line show linear fits and the dashed line represents the linear regression method with error implementation (see also Appendix A).*

be effective in the thermosphere and in the upper mesosphere. But, by using HALOE data, only a NO increase at altitudes above 80 km can be observed and lower altitudes do not seem to be affected lower in contrast to SPE.

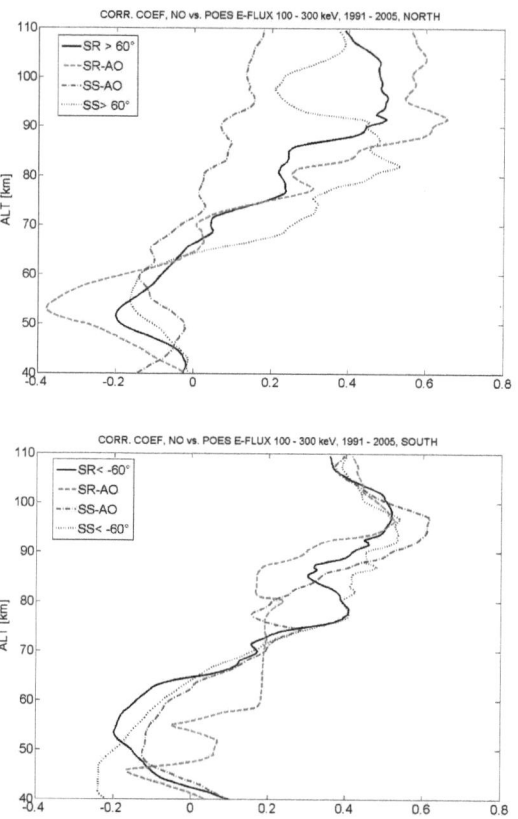

Figure 7.10: *Correlation coefficients of NO and electron flux 100 - 300 keV from 1991 - 2005 for northern hemisphere (upper panel) and southern hemisphere (lower panel). The solid line shows sunrise data at latitudes greater than 60°, dashed line data measured in the auroral oval in sunrise mode. The dashed dotted line monitors data in sunset mode in the auroral oval and the dotted line shows sunset data at latitudes greater than 60°.*

7.3 Indirect Effects (IE)

During polar winter and thus polar night conditions, the lifetime of NO_x is long enough throughout the mesosphere and lower thermosphere, that there is time enough for NO_x to descend to the stratosphere to participate in catalytic cycles controlling ozone. This coupling process of the upper and lower atmosphere is referred to as EEP indirect effect (IE) and was also investigated and discussed by Randall et al. [2007] and references therein. For this, it was important to distinguish between the NO_x increase caused by SPEs and this increase caused by EEPs. Data were excluded during days where high proton fluxes were detected to assure all effects of NO_x increase are mainly caused by electron precipitation. To obtain a good correlation between electron flux at polar regions and NO_x [vmr] values we focused on data from HALOE looking at the winter months for each hemisphere. In case of the northern hemisphere the months November, December, January and February (N, D, J, F) and for the southern hemisphere May, June, July and August (M, J, J, A) were taken into account. In these months the transport from the lower thermosphere down into the mesosphere and stratosphere is most effective. Due to atmospheric vertical dynamical processes the very stable species descending at polar regions to the middle mesosphere and stratosphere with about 1200 m/day during polar winter [Funke et al., 2005, 2008] leading to an increase of NO_x at low altitudes.

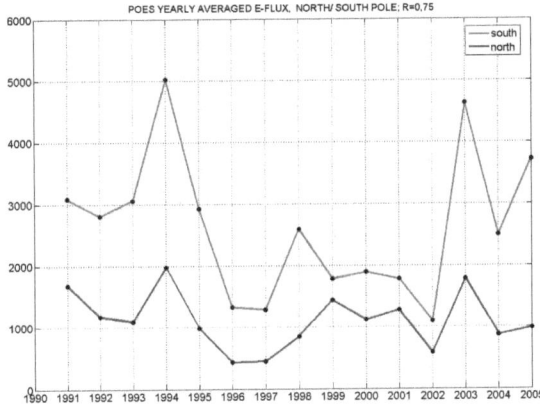

Figure 7.11: *Yearly averaged POES electron flux (300 keV - 2.5 MeV) covering the years 1991 to 2005, separated in North and South hemisphere. Data derived and obtained from J. M. Wissing, University of Osnabrück.*

In Fig. 7.12 these yearly averaged winter values for the southern hemisphere are shown, as well as the minimum and maximum values measured during the winter months. The investigation of the southern hemisphere shows more reliable results due to fact, that the big SPEs in 1994 and especially in 2003 did not occur during southern hemispheric wintertime. Also studies, carried out by Randall et al. [2007] during southern hemispheric wintertime, are in a good agreement with our results. Hence, if we look at Fig. 7.11 the yearly averaged POES electron flux calculated for the southern and northern hemisphere, we can observe that first the averaged values for the southern hemisphere seem to be always higher compared to the northern hemisphere, which could be a consequence of the South atlantic anomaly, which would lead to an increase of energetic electron flux. Another significant point shown in Fig. 7.11 are the high fluxes in the years 1994 and 2003. The increased values in 2003 are surely the matter of the period of high solar activity during these years leading to a high impact of charged particles, and of course also of electrons at the polar regions. But the year 1994 shows the interesting effect of increased electron flux values during the minimum of the solar cycle. This phenomenon has been observed before and is also discussed in section 3.2.2. Now comparing these fluxes to the results in Fig. 7.12 we also obtain during the years 1994 and 2003 very high averaged values of NO during the winter months in both hemispheres. Yearly averaged values of NO (during SH winter months) at an altitude of 70 km show 33 ppb in 1994 and 27 ppb in 2003 compared to background values of < 15 ppb in normal years. It has to be mentioned that the high values e.g. of the year 2003 in the southern hemisphere only show the effects of the precipitating electrons and not of the large SPE in 2003, since this happened months later in October and not during SH winter. Increased NO values e.g in 1994 and 2003 can be found from roughly 90 km down to about 40 km. Looking at correlation coefficients between NO values and the different sources of electron flux measurements and the Ap-index we find a very good correlation of NO vmr with the POES electron flux of $R = 0.88$ and also for the comparison of NO vmr with the AP index of $R = 0.81$ at 70 km altitude. Therefore a IE effect on the NO species seem to be very significant throughout the mesosphere and is investigated further in section 7.3.2.

7.3. Indirect Effects (IE)

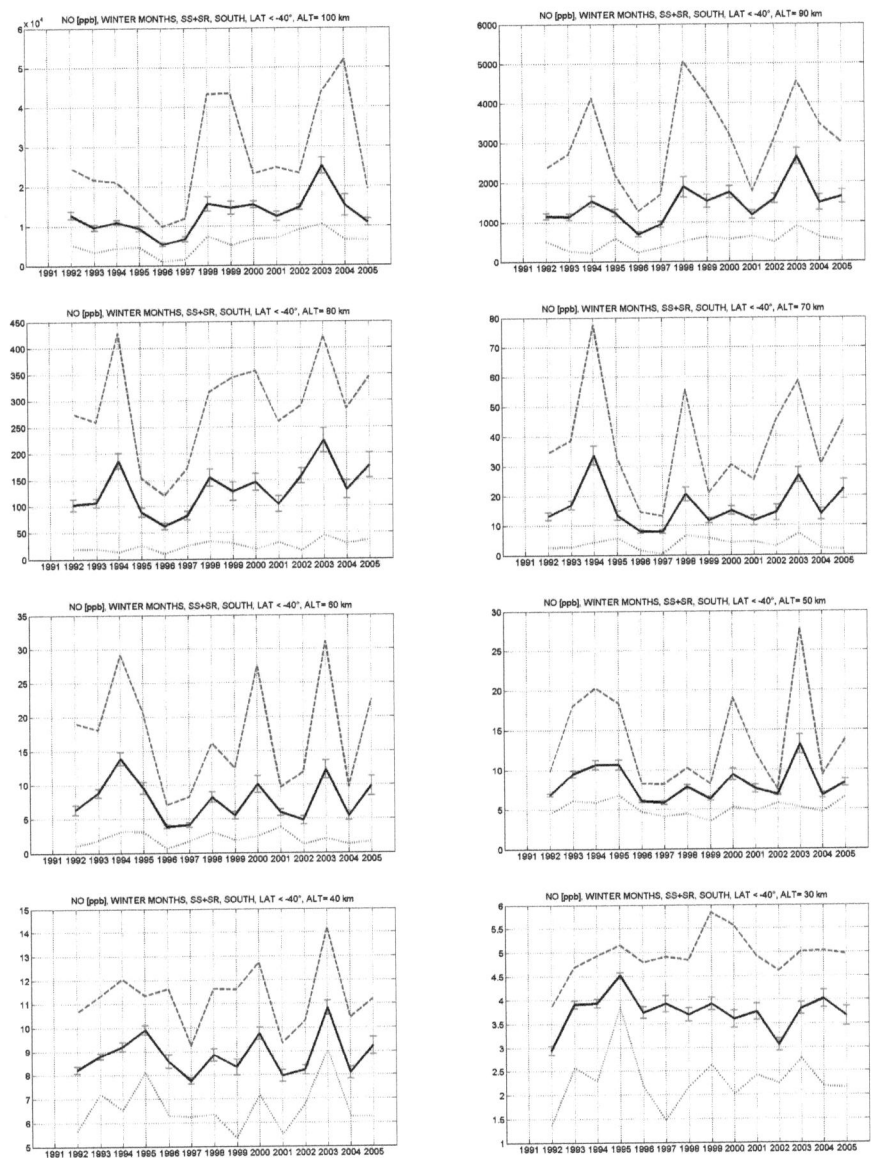

Figure 7.12: NO [ppb] mean values for southern hemispheric winter covering the years 1991 to 2005 at latitudes poleward of $40°$ south. Black solid line shows averaged values of whole winter NO in specific year, dashed line shows maximum values and the dotted line monitors the minimum values.

7.3.1 Indirect Effects (IE) and NO_x correlation

In section. 7.3 the effects of the IE on the NO budget during the winter months were already mentioned and its penetration range between roughly 90 and 40 km was briefly discussed. To obtain more precise statements about the effectiveness and penetration depth of the IE, correlation coefficients from NO data during the winter months and the POES, GOES electron fluxes and the Ap-index at the same time were calculated, covering the years 1991 to 2005 for both hemispheres.

Fig. 7.13 shows calculated correlation coefficients of yearly averaged NO in wintertime from HALOE data for the years 1991 to 2005 at altitudes from 35 to 100 km with GOES (R_G) and POES (R_P) electron fluxes and AP - index (R_{AP}) at the same times for the Northern hemisphere (upper panel) and the Southern hemisphere (lower panel). Results for the SH show the strongest correlation for POES electron flux with its maximum peak between 60 and 70 km altitude of $R_P = 0.92$. At this altitude also the Ap-index and the GOES electrons show their maximum correlation with $R_G \approx 0.6$ and $R_{AP} = 0.85$. Looking at the results in Fig. 7.13 (lower panel) for the SH, the Ap-index and the POES electron flux lead to a good agreement with the NO vmr values at altitudes from about 45 - 80 km with correlation coefficients larger than R = 0.7. The GOES electron flux in contrast does not seem to correlate very well with NO in the SH. This situation is quite different in the Northern hemisphere (see Fig. 7.13, upper panel). The maximum correlation is also given between 60 and 80 km, but in contrast to the SH, in the NH the GOES electrons and the Ap-index show the best correlation with $R_G \approx 0.82$ and $R_{AP} \approx 0.85$. At altitudes below 60 km and above 80 km the correlation, except for GOES, drops below 0.5. In general the shape of the correlation coefficients is more inconsistent in the NH in contrast to the SH. To investigate changes in $NO_x =$ (NO + NO_2), the NO_2 values, measured only to about 50 km altitude, were added to the NO data and the coefficients were calculated again. The new calculation showed no significant changes in the correlation coefficients, thus the plots are not shown.

Further, calculation of the correlation coefficients for lower energy ranges 100 - 300 keV were carried out concerning the IE, see Fig. 7.14. In case of the SH we obtain also very high coefficients > 0.8 for altitudes between roughly 45 and 75 km, comparable to the results calculated with higher energetic electron channels in Fig. 7.13. In the NH the correlation coefficients show very low values calculated for the lower energetic electrons. Between 60 and 80 km the coefficient reaches values > 0.5, only at higher altitudes at about 90 km we can observe a strong increase of the coefficient.

7.3.2 Indirect Effects (IE) and O_3 correlation

NO_x is among the other compounds created by precipitating electrons in the mesosphere and lower thermosphere and keeps stable during polar winter due to low dissociation rates. Thus, it can further descend to the stratosphere where it leads to ozone depletion by the catalytic NO_x cycle [Brasseur et al., 1999, Lary, 1997]. As the NO_x cycle is very dominant in the stratosphere we expected a strong anti correlation between NO_x and O_3 in the altitudes of 20-50 km. This effect is proved by HALOE data, which provides ozone data at altitudes from 15 to about 70 km. In Fig. 7.15 (upper panel) the

7.3. Indirect Effects (IE)

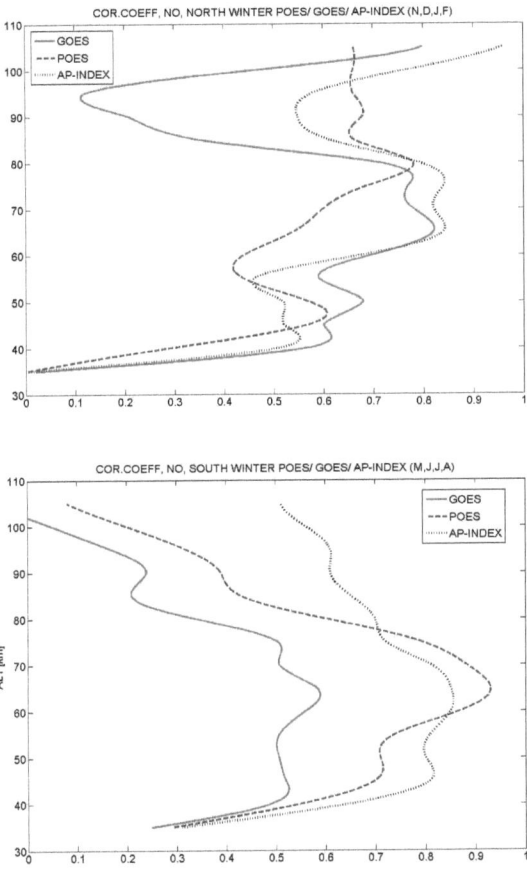

Figure 7.13: *Correlation coefficients derived from yearly averaged HALOE NO data (1991 - 2005) for northern hemispheric winter months (upper panel) and southern hemispheric winter months (lower panel) with GOES (E > 2 MeV), POES (300 keV - 2.5 MeV) and Ap-index data. Only HALOE data measured between latitudes from $40°$-$90°$ have been taken into account.*

correlation between yearly averaged wintertime NO, NO_x and O_3 is shown for latitudes from $40°$ to $90°$. As can be seen in the NH, the strongest effects are found between 28 km and 30 km reaching a maximum of R ≈ -0.9. Assuming only values smaller than

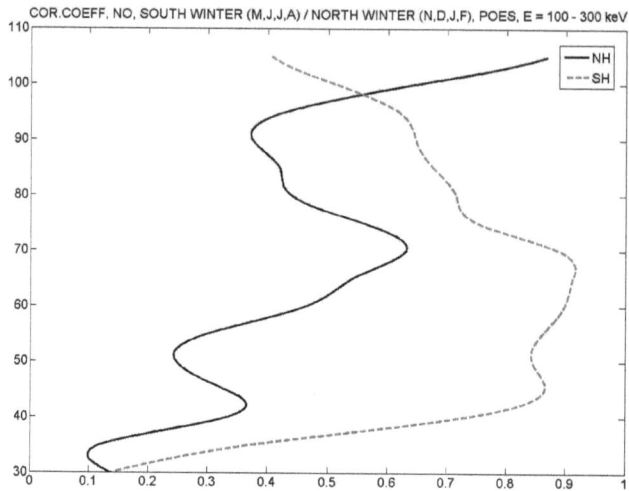

Figure 7.14: *Correlation coefficients derived from yearly averaged HALOE NO data (1991 - 2005) for northern hemispheric winter months (solid line) and southern hemispheric winter months (dashed line) with POES electron fluxes (100 keV - 300 keV). Only HALOE data measured between latitudes from N/S 40° 90° have been taken into account.*

-0.6 are a strong result of anti-correlation, the maximum ozone loss cycle depending on NO_x seems to occur only at altitudes between 25 and 35 km during polar winter. At other altitudes, several other cycles seem to force the ozone depletion. In case of the northern hemisphere, hardly any differences between the correlation coefficients of NO and NO_x could be found. NO as well as NO_2 seem to be affected the same way.

In the southern hemisphere (Fig. 7.15 lower panel) the shape of the NO_x/O_3 correlation is slightly different. Again assuming the coefficient smaller than -0.6 show a strong linear correlation of the NO_x cycle, we can observe two peaks. The maximum peak of anti-correlation is given for altitudes from 20 to 30 km reaching maximum values R = -0.8, and seems to be a little shifted in contrast to the NH. Above this altitude of about 30-40 km, hardly any correlation can be seen, thus the NO_x and ozone values are not correlated at all. This might be an indication of the very important ClO_x cycle becoming a dominating process at these altitudes competing with the NO_x catalytic cycle. As an interesting feature a second peak can be observed between 50 and 60 km reaching a peak value of -0.75. This might be an indication of a somehow shifted transport process at these altitudes in the southern hemisphere. Also important to be noted, in the SH the correlation NO_x/O_3 and only NO/O_3 differs significantly at

7.3. Indirect Effects (IE)

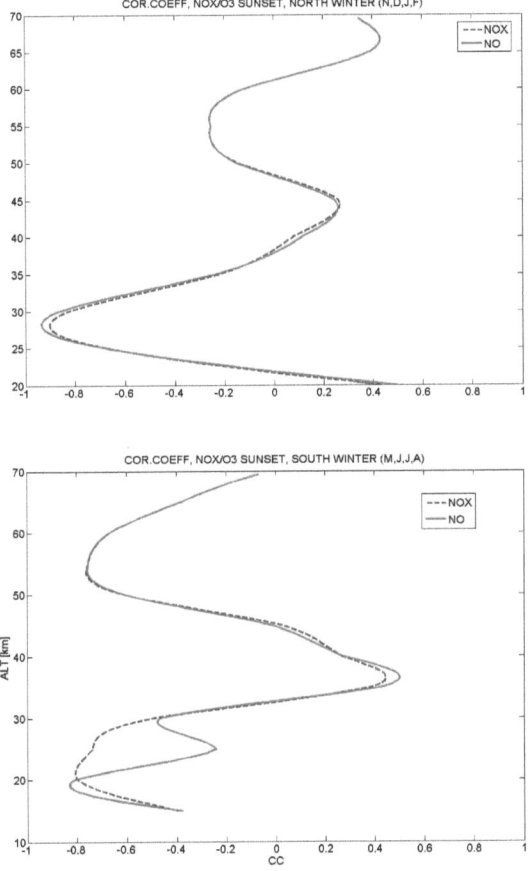

Figure 7.15: *Correlation coefficients derived from yearly averaged HALOE O_3 with NO (solid line) and NO_x, data (1991 - 2005) for northern hemispheric winter months (upper panel) and southern hemispheric winter months (lower panel). Only HALOE data measured between latitudes from N/S 40°-90° have been taken into account.*

altitudes between 20 and 30 km. This might be caused due to the strong formation of the 'ozone hole' at the south pole which consumes large amounts of NO and thus leads to an unclear correlation at this altitudes. Hence, NO and NO_2 have to be processed together to monitor the ozone depletion due to the NO_x cycle.

As NO and NO_2 are converted during nighttime into species like e.g. HNO_3, a correlation between NO_x/O_3 can only be measured during daytime. Also the HALOE data only show a good anti correlation of NO_x/O_3 in the sunset measurements. Measurements of sun rise data showed no anti-correlation at all.

7.4 Discussion EEP

Comparisons of several instruments measuring highly energetic electron fluxes showed that POES data and the Ap-index fit very well together, if they are monthly averaged. The Ap-index is a good proxy for electron fluxes entering the polar regions.
Another interesting point can be seen by comparing the Ap-index correlation plot (SH) and the POES low energetic electron plot (SH). Both show nearly about the same shape and the same high values. Thus, the Ap-index once again seems to be a good proxy for electron fluxes in the keV energy range. The study also showed that the EEP direct effect has a significant effect on NO_x in the upper atmosphere and leads to an increase of NO_x down to altitudes of 80 km but not below. Long term studies of the NO HALOE data set during wintertime for the northern hemisphere and southern hemisphere showed no correlation of the EEP direct effect with electrons in the energy ranges from 300 keV - 2.5 MeV, but a correlation of up to 0.6 at roughly 90 to 100 km in regions of the auroral oval (Lat = 55° - 65°). A lower correlation coefficient of 0.5 can be found at 80 km, thus confirming our result from the single EEP, mentioned before, which also shows an increase of NO down to altitudes of about 80 km. The differences between the correlation profiles (see Fig. 7.10) of the NH and SH is not quite clear. Either it is evidence for a different particle influx or it can also be a consequence of too little data points available for the polar regions.
In case of the indirect effect (IE) we obtained very high correlation coefficients (see Fig. 7.13) for the northern and southern hemisphere with electron fluxes with energies ranging from 300 keV - 2.5 MeV as well as with the lower energy channel from 100 keV - 300 keV (see Fig. 7.14). In the SH both energy groups and the POES data as well as the Ap-index show a distinct maximum of high correlation coefficients between roughly 40 and 80 km . Results show that the transport processes in southern hemisphere seem to be much stronger and more stable in contrast to the northern hemisphere. The coefficients obtained from the lower energy channels might also give somehow evidence that the POES/MEPED instrument, as mentioned in section 7.1, is most sensitive to electrons with 300 keV - 600 keV. Thus, both electron channels contain mainly data of 300 keV and therefore both show a similar correlation. This confirms previous studies by Randall et al. [2007], which also show strong correlation between NO_x and enhanced highly energetic electron fluxes caused by EEPs in the southern hemisphere and which descends down to altitudes of 45 km each winter, but have not considered the northern hemisphere.
It also strengthens the hypothesis of Sinnhuber et al. [2005] that large decadal scale O_3 variations in the northern hemispheric polar winter correlated to GOES electron fluxes could be a result of EEPs. This study also confirmed a large correlation of

7.4. Discussion EEP

GOES electron fluxes in the NH and NO. However, it should be pointed out that the NO_x enhancements studied here would reach the mid-stratosphere only at the end of the winter, thus they can not explain the large O_3 variability in high winter in the mid-stratosphere.

8. Summary and Conclusion

The long term data set, consisting of measurements from the years 1991 to 2005, from the HALOE instrument onboard the UARS satellite has been investigated regarding effects of highly energetic particle events on the chemical composition of e.g. NO, NO_2, O_3 and HCl. In terms of energetic particle events we can distinguish between Solar Proton Events (SPE), characterized by a strong and impulsive increase of proton flux originated from massive solar eruptions with energies of several MeV, and electron precipitation events (EEP) determined by a sudden increase of highly energetic particles with energies in the order of keV to MeV. As the Earth and its atmosphere is shielded by the magnetosphere, protecting the environment from charged particles and solar radiation, highly energetic electrons and protons can only enter at polar regions, where the magnetic field lines are open and the particles can penetrate deep into the atmosphere.
In case of the SPEs, a clear increase of $NO_x = (NO + NO_2)$ in the middle atmosphere was observed, e.g during the Bastille Event in July 2000 and the Halloween Event October/November 2003, down to altitudes of roughly 45 km.
Also, a significant decrease of ozone in the middle and upper stratosphere due to the increased NO_x amounts during the SPE was be monitored by HALOE data.
During the Bastille Event, HCl, which is in general an inactive reservoir of the very reactive and ozone destroying species chlorine, was found to be slightly decreased. This decrease was observed the first time by this study. Comparison of the measurements with the UBIC model were in a good agreement and showed that dynamical as well as the chemicals effects during a SPE are very well understood.
SPEs were also found to affect the temperature of the atmosphere during the impact of the highly energetic protons. An increase of temperature of up to 180° during the Bastille event could be observed in the lower thermosphere and upper mesosphere. Due to the ozone depletion in the stratosphere during the SPE, the effect of a cooling process was expected. But HALOE data showed no strong evidence for this effect below $\approx$ 80 km. Thus a significant temperature decrease of the stratosphere caused by SPEs as a matter of lower O_3 amounts in the 'ozone layer' can not be confirmed using HALOE data, but not excluded either, because of the very difficult detection of these temperature changes in the middle and upper atmosphere.
The comparison of monthly POES electron flux data and the monthly Ap-index showed that those types of data fit very well and both appear to be a good proxy for long term highly energetic electron precipitation investigations.
It was shown that EEP direct effect leads to an increase of NO_x down to altitudes of

about 80 km, but does not seem to affect the chemical budget at altitudes below.
More effective is the observed EEP indirect effect (IE) which influences the year to year amounts of NO_x during the winter months on both hemispheres. Of course, this enhanced wintertime NO_x amounts affect directly the stratospheric ozone budget and lead to an decrease of stratospheric ozone during the years of high indirect highly electron precipitation events, which partially confirms results of previous studies by Sinnhuber et al. [2005] of a large decadal change of polar winter ozone due to the influence of highly energetic particle precipitation.

This study clearly showed that the impact of highly energetic particles events has a strong influence on the middle atmospheric chemistry and points out the differences between the highly energetic proton events (SPE) and highly energetic electron precipitation events (EEP). Both, the direct effect and the indirect effects have to be taken to account to improve the understanding of the atmospheric chemistry and dynamics, which is so far not completely understood.

A. Appendix

Earth's Magnetic Field

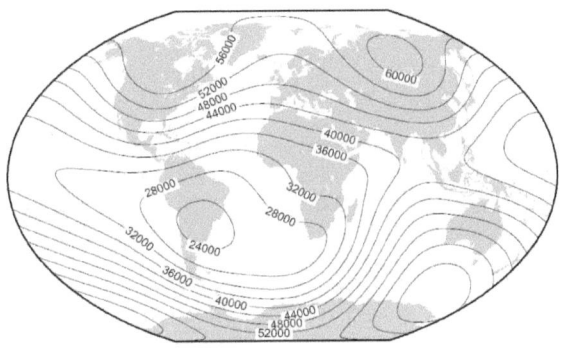

Figure A.1: *Contour maps of total intensity (nT) at the surface, adopted from the British Geological Survey, see:http://www.geomag.bgs.ac.uk.*

Constants

R_{Sun}	695,700 km
R_{Earth}	6,370 km
1 AU	149,597,870,691 ± 6 m
m_{e^-}	$9{,}1093897 \cdot 10^{-31}$ kg
m_p	$1{,}6726231 \cdot 10^{-27}$ kg
m_n	$1{,}6749286 \cdot 10^{-27} kg$
c	299 792 458 ms^{-1}
1 pfu	particle cm^{-2} s sr keV

Useful Tables

Energy	Wavelength	Frequency
12.4 feV	100 Mm	3 Hz
124 feV	10 Mm	30 Hz
1.24 peV	1 Mm	300 Hz
12.4 peV	100 km	3 kHz
124 peV	10 km	30 kHz
1.24 neV	1 km	300 kHz
12.4 neV	100 m	3 MHz
124 neV	10 m	30 MHz
1.24 μeV	1 m	300 MHz
12.4 μeV	1 dm	3 GHz
124 μeV	1 cm	30 GHz
1.24 meV	1 mm	300 GHz
12.4 meV	100 μm	3 THz
124 meV	10 μm	30 THz
1.24 eV	1 μm	300 THz
12.4 eV	100 nm	3 PHz
124 eV	10 nm	30 PHz
1.24 keV	1 nm	300 PHz
12.4 keV	100 pm	3 EHz
124 keV	10 pm	30 EHz
1.24 MeV	1 pm	300 EHZ

Table A.15: *Tabular of the electromagnetic spectrum in terms of energy, wavelength and frequency.*

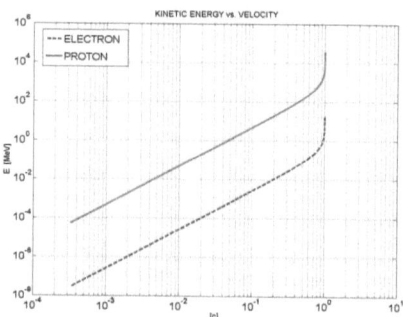

Figure A.2: *Relativistic E_{Kin} versus velocity in [c] of protons (solid line) and electrons (dashed line).*

Linear Regression

In case of data analysis, very often the question t has to be solved wether a set of independent variables is related to each other or no. Thus, to get a statistical predication, the method of linear regression has to be used (see Fig. A.3), to get a quantitative statement for a given set of data.

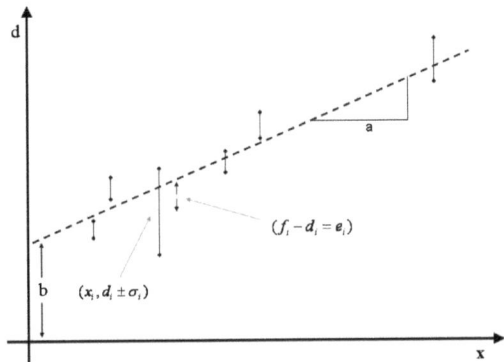

Figure A.3: *Sketch of linear regression problem.*

This method uses the least squares function to find a linear correlation, expressed by the linear regression function, for a group of data points (see also Lang and Pucker [2005]). In the special form of linear regression used in our problem the quality of the data set is also included and the error of the measurements are implied into the calculation of the regression equation and its coefficients. In case of this method the function **f** for a given set of data points d_i and its error $\pm \sigma_i$ at the positions x_i (i=1,n) has to be found representing the best appropriate solution.

$$f = a \cdot x + b \tag{A.56}$$

As the error σ_i is included, the calculated coefficients **a** and **b** will be influenced and can be described by

$$F(a,b) = \sum_{i=1}^{n} \left(\frac{e_i}{\sigma_i}\right)^2 = \sum_{i=1}^{n} \left[\frac{(ax_i + b - d_i)^2}{\sigma_i^2}\right] \tag{A.57}$$

Solving this equation after the coefficients where the function **F(a,b)** is set to minimum

$$\frac{\partial F}{\partial a}\Big|_{a,b} = 0 \longrightarrow min$$
$$\frac{\partial F}{\partial b}\Big|_{a,b} = 0 \longrightarrow min \tag{A.58}$$

These equations lead to the terms

$$\alpha = \sum\left(\frac{x_i^2}{\sigma_i^2}\right), \ \beta = \sum\left(\frac{x_i}{\sigma_i^2}\right), \ \gamma = \sum\left(\frac{x_i d_i}{\sigma_i^2}\right),$$
$$\delta = \sum\left(\frac{d_i}{\sigma_i^2}\right), \ \epsilon = \sum\left(\frac{1}{\sigma_i^2}\right) \tag{A.59}$$

which further build the equation system with our new coefficients $\hat{a}$ and $\hat{b}$

$$\alpha \cdot \hat{a} + \beta \cdot \hat{b} = \gamma$$
$$\beta \cdot \hat{a} + \epsilon \cdot \hat{b} = \delta \tag{A.60}$$

By solving this equation system we can calculate our new coefficients with implied data error by

$$\hat{b} = \frac{\alpha \cdot \delta - \beta \cdot \gamma}{\alpha \cdot \epsilon - \beta^2}$$
$$\hat{a} = \frac{\gamma - \beta \cdot \hat{b}}{\alpha} \tag{A.61}$$

which finally derives our new fitting function (see Fig.A.3) with the implied error by

$$\hat{f} = \hat{a} \cdot x + \hat{b} \tag{A.62}$$

This method has been used e.g. in chapter. 6.1.4 to obtain the best solution for the linear correlation in terms of long time direct energetic electron precipitation effects and its impact on the middle atmosphere's NO_x budget.

Classification of Solar Radio Bursts

- **Type I** - Short, narrow band events that usually occur in great numbers together with a broader band continuum, duration ≈ hours, days.

- **Type II** - Slow drift from high to low frequencies. Often show fundamental and second harmonic frequency structure.

- **Type III** - Rapidly drift from high to low frequencies. May exhibit harmonics. Often accompany the flash phase of large flares.

- **Type IV** - Flare-related broad-band continua

- **Type V** - Broad-band continua which may appear with III bursts, duration ≈ minutes, $\tau \sim \frac{1}{f}$.

B. References

Publications related to the thesis

Kazeminejad, S., Sinnhuber, M., Notholt, J., Winkler, H., Wissing, J.M., *Effects of the high energetic electron precipitation events at polar regions of the middle atmosphere.*, Journal of Geophysical Research, 2009, (under submission)

Winkler, H., **Kazeminejad, S.**, Sinnhuber, M., Notholt, *Conversion of mesospheric HCl into active chlorine during the solar proton event in July 2000 in the northern polar region.*, Journal of Geophysical Research, 2008, (under submission)

Conference contributions

Kazeminejad, S., Sinnhuber, M., Notholt, J., von Savigny, C., *Analysis of NO_x in the middle atmosphere determined by HALOE*, Deutsche Physikalische Gesellschaft, DPG - Tagung, Heidelberg, March 2006

Kazeminejad, S., Sinnhuber, M., Notholt, J., von Savigny, C., *Analysis of NO_x in the middle atmosphere determined by HALOE*, EGU General Assembly, Vienna, April 2006

Kazeminejad, S., Sinnhuber, M., Notholt, J., Winkler, H., *Response of the middle atmospheres chemical composition due to Solar Particle Events*, AGU fall meeting, San Francisco, December 2007

Kazeminejad, S., Sinnhuber, M., Notholt, J., Winkler, H., *Response of the middle atmospheres chemical composition due to Solar Particle Events*, EGU General Assembly, Vienna, April 2008

Kazeminejad, S., Sinnhuber, M., Notholt, J., Winkler, H., *Influence of Particle Events on the Middle Atmosphere*, FMI - HEPPA Workshop, Helsinki, May 2008

Bibliography

S.-I. Akasofu and S. Chapman. *Solar-terrestrial physics. an account of the wave and particle radiations from the quiet and the active sun, and of the consequent terrestrial phenomena.* The International Series of Monographs on Physics, Oxford: Clarendon Press,, 1972.

H. Alfven. Electric currents in cosmic plasmas. *Reviews of Geophysics and Space Physics*, 15:271–284, August 1977.

D. G. Andrews. *An Introduction to Atmospheric Physics.* An Introduction to Atmospheric Physics, by David G. Andrews, pp. 240. ISBN 0521620511. Cambridge, UK: Cambridge University Press, 2000.

M. Aschwanden. *Physics of the Solar Corona. An Introduction with Problems and Solutions* . Springer Verlag, Berlin; Volume: 1. A. , ISBN 978-3540307655, 2006.

D. N. Baker, R. W. Klebesadel, P. R. Higbie, and J. B. Blake. Highly relativistic electrons in the earth's outer magnetosphere. I - Lifetimes and temporal history 1979-1984. *Journal of Geophysical Research, Vol.91*, pages 4265–4276, April 1986.

S. J. Bauer and H. Lammer. *Planetary aeronomy : atmosphere environments in planetary systems.* S.J. Bauer and H. Lammer. Springer Verlag. ISBN 3-540-21472-0, 2004.

D. J. Bombardieri, M. L. Duldig, K. J. Michael, and J. E. Humble. Relativistic Proton Production during the 2000 July 14 Solar Event: The Case for Multiple Source Mechanisms. *The Astrophysical Journal*, 644:565–574, June 2006.

G. P. Brasseur, J. J. Orlando, and G. S. Tyndall. *Atmospheric Chemistry and Global Change.* Oxford University Press, USA; 1st edition (March 15, 1999) ISBN 0-19-510521-4, 1999.

G. P. Brasseur and S. Solomon. *Aeronomy of the Middle Atmosphere: Chemistry and Physics of the Stratosphere and Mesosphere.* G.P. Brasseur and S. Solomon. 2005 XII, 644 p. 3rd rev. and enlarged ed. 1-4020-3284-6. Berlin: Springer, 2005.

L. B. Callis, R. E. Boughner, D. N. Baker, J. B. Blake, and J. D. Lambeth. Precipitating relativistic electrons - Their long-term effect on stratospheric odd nitrogen levels. *Journal of Geophysical Research, Vol.96*, pages 2939–2976, February 1991.

L. B. Callis, M. Natarajan, D. S. Evans, and J. D. Lambeth. Solar - atmospheric coupling by electrons (SOLACE). 1. Effects of the May 12, 1997 solar event on the middle atmosphere. *Journal of Geophysical Research*, 103:28405–28419, November 1998.

R. C. Carrington. Description of a Singular Appearance seen in the Sun on September 1, 1859. *Monthly Notices of the Royal Astronomical Society*, 20:13–15, November 1859.

J. W. Chamberlain. Planetary coronae and atmospheric evaporation. *Planetary Space Science*, 11:901–+, August 1963.

S. Chapman. On ozone and atomic oxygen in the upper atmosphere. *Philosophical Magazine*, 7:369–383, July 1930.

M. P. Chipperfield. Multiannual simulations with a three-dimensional chemical transport model. *Journal of Geophysical Research*, 104:1781–1806, 1999.

M. Clilverd. Remote sensing space weather events: the AARDDVARK network. *Talk, High-Energy Particle Precipitation in the Atmosphere (HEPPA) meeting 2008, FMI - Finland*, 2008.

H. Cremades and V. Bothmer. On the three-dimensional configuration of coronal mass ejections. *Astronomy and Astrophysics*, 422:307–322, July 2004.

P. J. Crutzen, I. S. A. Isaksen, and G. C. Reid. Solar Proton Events: Stratospheric Sources of Nitric Oxide. *Science*, 189:457–459, August 1975.

D. A. Degenstein, R. L. Gattinger, N. D. Lloyd, A. E. Bourassa, J. T. Wiensz, and E. J. Llewellyn. Observations of an extended mesospheric tertiary ozone peak. *Journal of Atmospheric and Solar-Terrestrial Physics*, 67:1395–1402, October 2005.

R. A. Frahm, J. D. Winningham, J. R. Sharber, R. Link, G. Crowley, E. E. Gaines, D. L. Chenette, B. J. Anderson, and T. A. Potemra. The diffuse aurora: A significant source of ionization in the middle atmosphere. *Journal of Geophysical Research, Vol.102*, pages 28203–28214, December 1997.

J. Fritzenwallner and E. Kopp. Model calculations of the negative ion chemistry in the mesosphere with special emphasis on the chlorine species and the formation of cluster ions. *Advances in Space Research*, 21:891–894, 1998.

B. Funke, M. López-Puertas, S. Gil-López, T. von Clarmann, G. P. Stiller, H. Fischer, and S. Kellmann. Downward transport of upper atmospheric NO_x into the polar stratosphere and lower mesosphere during the Antarctic 2003 and Arctic 2002/2003 winters. *Journal of Geophysical Research (Atmospheres)*, 110:24308–+, December 2005. doi: 10.1029/2005JD006463.

B. Funke, M. López-Puertas, S. Gil-López, T. von Clarmann, G. P. Stiller, H. Fischer, and S. Kellmann. Energetic electron precipitation effects on the polar winter stratosphere as observed by MIPAS/Envisat . *Talk, HEPPA MEETING, Helsinki*, May 2008.

E. E. Gaines, D. L. Chenette, W. L. Imhof, C. H. Jackman, and J. D. Winningham. Relativistic electron fluxes in May 1992 and their effect on the middle atmosphere. *Journal of Geophysical Research, Vol.100*, pages 1027–1033, January 1995.

K. H. Glassmeier and M. Scholer. *Plasmaphysik im Sonnensystem*. BI-Wiss. Verlag, Mannheim, ISBN 3-411-15151-X, 1991.

L. L. Gordley, J. M. Russell, III, L. J. Mickley, J. E. Frederick, J. H. Park, K. A. Stone, G. M. Beaver, J. M. McInerney, L. E. Deaver, G. C. Toon, F. J. Murcray, R. D. Blatherwick, M. R. Gunson, J. P. D. Abbatt, R. L. Mauldin, III, and et al. Validation of nitric oxide and nitrogen dioxide measurements made by the Halogen Occultation Experiment for UARS platform. *Journal of Geophysical Research*, 101: 10240–10266, April 1996.

C. H. Jackman, M. T. Deland, G. J. Labow, E. L. Fleming, D. K. Weisenstein, M. K. W. Ko, M. Sinnhuber, J. Anderson, and J. M. Russell. The influence of the several very large solar proton events in years 2000 - 2003 on the neutral middle atmosphere. *Advances in Space Research*, 35:445–450, 2005a.

C. H. Jackman, M. T. DeLand, G. J. Labow, E. L. Fleming, D. K. Weisenstein, M. K. W. Ko, M. Sinnhuber, and J. M. Russell. Neutral atmospheric influences of the solar proton events in October-November 2003. *Journal of Geophysical Research (Space Physics)*, 110:9–+, July 2005b.

C. H. Jackman, E. L. Fleming, and F. M. Vitt. Influence of extremely large solar proton events in a changing stratosphere. *J. Geophys. Res.*, 105:11659–11670, May 2000.

C. H. Jackman, R. D. McPeters, G. J. Labow, E. L. Fleming, C. J. Praderas, and J. M. Russell. Northern hemisphere atmospheric effects due to the July 2000 solar proton event. *Geophysikal Research Letters*, 28:2883–2886, August 2001.

C. H. Jackman, R. G. Roble, and E. L. Fleming. Mesospheric dynamical changes induced by the solar proton events in October-November 2003. *Geophysical Research Letters,*, 34:4812–+, February 2007.

C. Kalicinsky. Prezipierende magnetosphrische Teilchen und die Zuverlaessigkeit der Satellitenmessungen. Master's thesis, University Osnabrueck, Fachbereich Physik, Arbeitsgruppe Modellierung, 2008.

S. Kazeminejad. Exospheric Temperature Estimation and Atmospheric Loss: a Comparative Study of Mars and Venus. Master's thesis, AA(Karl-Franzens-University Graz, Institute for Physics, Universitaetsplatz 5, A-8010 Graz, Austria), 2005.

C. U. Keller, J. O. Stenflo, S. K. Solanki, T. D. Tarbell, and A. M. Title. Solar magnetic field strength determinations from high spatial resolution filtergrams. *Astronomy and Astrophysics*, 236:250–255, September 1990.

J. S. Kinnersley. The climatology of the stratospheric THIN AIR model. *Quarterly Journal of the Royal Meteorological Society*, 122:219–252, January 1996.

M. G. Kivelson and C. T. Russel. *Introduction to space physics*. Cambridge University Press, 1995.

A. A. Krivolutsky, A. V. Klyuchnikova, G. R. Zakharov, T. Y. Vyushkova, and A. A. Kuminov. Dynamical response of the middle atmosphere to solar proton event of July 2000: Three-dimensional model simulations. *Advances in Space Research*, 37: 1602–1613, 2006.

S. Krucker and R. P. Lin. Relative Timing and Spectra of Solar Flare Hard X-ray Sources. *Solar Physics*, 210:229–243, November 2002.

C. B. Lang and N. N. Pucker. *Mathematische Methoden in der Physik*. Elsevier - Spektrum Akademischer Verlag, Heidelberg-Berlin ISBN 3-8274-1558-6, 2005.

D. J. Lary. Catalytic destruction of stratospheric ozone. *Journal of Chemical Physics*, 102:21515–21526, 1997.

R. P. Lin. Relationship of solar flare accelerated particles to solar energetic particles (SEPs) observed in the interplanetary medium. *Advances in Space Research*, 35: 1857–1863, 2005.

M. López-Puertas, B. Funke, S. Gil-López, T. von Clarmann, G. P. Stiller, M. Höpfner, S. Kellmann, G. Mengistu Tsidu, H. Fischer, and C. H. Jackman. $HNO_3, N_2O_5, and ClONO_2$ enhancements after the October-November 2003 solar proton events. *Journal of Geophysical Research (Space Physics)*, 110(9):9–+, September 2005.

R. McPherron. Predicting the Ap index from past behavior and solar wind velocity. *Physics and Chemistry of the Earth C*, 24:45–56, 1999. doi: 10.1016/S1464-1917(98)00006-3.

M. Menvielle and A. Berthelier. The K-derived planetary indices - Description and availability. *Reviews of Geophysics*, 29:415–432, August 1991.

R. A. Mewaldt, C. M. S. Cohen, A. W. Labrador, R. A. Leske, G. M. Mason, M. I. Desai, M. D. Looper, J. E. Mazur, R. S. Selesnick, and D. K. Haggerty. Proton, helium, and electron spectra during the large solar particle events of October-November 2003. *Journal of Geophysical Research (Space Physics)*, 110:9–+, September 2005.

J. A. Miller, P. J. Cargill, A. G. Emslie, G. D. Holman, B. R. Dennis, T. N. LaRosa, R. M. Winglee, S. G. Benka, and S. Tsuneta. Critical issues for understanding particle acceleration in impulsive solar flares. *Journal of Geophysical Research*, 102: 14631–14660, July 1997.

D. G. Murcray, J. Gillis, A. Goldman, and J. J. Kosters. *Stratospheric constituent measurements using UV solar occultation technique, Colorado University Report, Technical Report*. Nauenberg, U., 1981.

K. W. Ogilvie, D. J. Chornay, R. J. Fritzenreiter, F. Hunsaker, J. Keller, J. Lobell, G. Miller, J. D. Scudder, E. C. Sittler, Jr., R. B. Torbert, D. Bodet, G. Needell, A. J. Lazarus, J. T. Steinberg, J. H. Tappan, A. Mavretic, and E. Gergin. SWE, A Comprehensive Plasma Instrument for the Wind Spacecraft. *Space Science Reviews*, 71:55–77, February 1995. doi: 10.1007/BF00751326.

T. G. Onsager, G. Rostoker, H.-J. Kim, G. D. Reeves, T. Obara, H. J. Singer, and C. Smithtro. Radiation belt electron flux dropouts: Local time, radial, and particle-energy dependence. *Journal of Geophysical Research (Space Physics)*, 107:1382–+, November 2002.

H. S. Porter, C. H. Jackman, and A. E. S. Green. Efficiencies for production of atomic nitrogen and oxygen by relativistic proton impact in air. *Journal of Chemical Physics*, 65:154–167, July 1976.

T. Pulkkinen. Space Weather: The Terrestrial Perspective. *Living Reviews in Solar Physics*, 4:2–+, May 2007.

W. Raith, editor. *Bergmann–Schäfer – Lehrbuch der Experimentalphysik, ISBN: 3-11-016837-5*, volume 7 – Erde und Planeten. W.de Gruyter, Berlin, 1997.

C. E. Randall, V. L. Harvey, C. S. Singleton, S. M. Bailey, P. F. Bernath, M. Codrescu, H. Nakajima, and J. M. Russell. Energetic particle precipitation effects on the Southern Hemisphere stratosphere in 1992-2005. *Journal of Geophysical Research (Atmospheres), Vol.112*, pages 8308–+, April 2007.

G. C. Reid, S. Solomon, and R. R. Garcia. Response of the middle atmosphere to the solar proton events of August-December, 1989. *Geophysical Research Letters*, 18: 1019–1022, June 1991.

J. H. Reid. Classification of Solar Flares. *Irish Astronomical Journal*, 6:45–+, June 1963.

M. J. Reiner, M. L. Kaiser, M. Karlický, K. Jiřička, and J.-L. Bougeret. Bastille Day Event: A Radio Perspective. *Solar Physics*, 204:121–137, December 2001.

I. G. Richardson, E. W. Cliver, and H. V. Cane. Sources of geomagnetic activity over the solar cycle: Relative importance of coronal mass ejections, high-speed streams, and slow solar wind. *Journal of Geophysical Research (Space Physics)*, 105:18203–18214, 2000.

C. J. Rodger, M. A. Clilverd, N. R. Thomson, D. Nunn, and J. Lichtenberger. Lightning driven inner radiation belt energy deposition into the atmosphere: regional and global estimates. *Annales Geophysicae*, 23:3419–3430, December 2005.

C. J. Rodger, T. Raita, M. A. Clilverd, A. Seppälä, S. Dietrich, N. R. Thomson, and T. Ulich. Observations of relativistic electron precipitation from the radiation belts driven by EMIC waves. *Geophysical Research Letters,*, 35:16106–+, August 2008.

I. Roth and S. D. Bale. Heliospheric ion energization due to emerging CME shocks. *Journal of Geophysical Research (Space Physics)*, 111:7–+, July 2006.

D. W. Rusch, J.-C. Gérard, S. Solomon, P. J. Crutzen, and G. C. Reid. The effect of particle precipitation events on the neutral and ion chemistry of the middle atmosphere. *Planetary Space Science*, 29:767–774, July 1981.

C. T. Russell. The solar wind interaction with the Earth's magnetosphere: a tutorial. *IEEE Transactions on Plasma Science*, 28:1818–1830, December 2000. doi: 10.1109/27.902211.

J. M. Russell, III, L. L. Gordley, J. H. Park, S. R. Drayson, W. D. Hesketh, R. J. Cicerone, A. F. Tuck, J. E. Frederick, J. E. Harries, and P. J. Crutzen. The Halogen Occultation Experiment. *Journal of Geophysical Research (Atmospheres)*, 98:10777–+, June 1993.

M. Scharringhausen. *Mesospheric and Thermospheric Magnesium Species from Space.* M. Scharringhausen. Vdm Verlag Dr. Mller. ISBN 3-836-49164-8, 2008.

J. Schröter, B. Heber, F. Steinhilber, and M. B. Kallenrode. Energetic particles in the atmosphere: A Monte-carlo simulation. *Advances in Space Research*, 37:1597–1601, 2006.

S.H. Schwabe. Sonnen -Beobachtungen im Jahre 1843. *Astronomische Nachrichten*, 21: 233–236, January 1844.

R. Schwenn. *Large-Scale Structure of the Interplanetary Medium*, pages 99–+. Physics of the Inner Heliosphere I, 1990.

R. Schwenn. Space Weather: The Solar Perspective. *Living Reviews in Solar Physics*, 3:2–+, August 2006.

B.-M. Sinnhuber, P. von der Gathen, M. Sinnhuber, M. Rex, G. König-Langlo, and S. J. Oltmans. Large decadal scale changes of polar ozone suggest solar influence. *Atmospheric Chemistry and Physics Discussions*, 5:12103–12117, 2005.

S. Solomon and P. J. Crutzen. Analysis of the August 1972 solar proton event including chlorine chemistry. *Journal of Geophysical Research*, 86:1140–1146, 1981.

W. Swider and T. J. Keneshea. Decrease of ozone and atomic oxygen in the lower mesosphere during a PCA event. *Planetary and Space Science*, 21:1969–1973, November 1973.

B. T. Tsurutani, D. L. Judge, F. L. Guarnieri, P. Gangopadhyay, A. R. Jones, J. Nuttall, G. A. Zambon, L. Didkovsky, A. J. Mannucci, B. Iijima, R. R. Meier, T. J. Immel, T. N. Woods, S. Prasad, L. Floyd, J. Huba, S. C. Solomon, P. Straus, and R. Viereck. The October 28, 2003 extreme EUV solar flare and resultant extreme ionospheric effects: Comparison to other Halloween events and the Bastille Day event. *Geophysical Research Letters*, 32:3–+, January 2005.

L. I. Vagina and V. A. Popov. Comparison of Acceleration Regions of Energetic Particles Measured at Low Latitudes with Locations of Substorm Wedges. In A. Wilson, editor, *Fifth International Conference on Substorms*, volume 443 of *ESA Special Publication*, pages 595–+, July 2000.

F. M. Vitt and C. H. Jackman. A comparison of sources of odd nitrogen production from 1974 through 1993 in the Earth's middle atmosphere as calculated using a two-dimensional model. *Journal of Geophysical Research (Space Physics)*, 101:6729–6740, 1996.

J. Vogt, M. Sinnhuber, and MB. Kallenrode. *Effects of Geomagnetic Variations on System Earth*. Verlag Springer Berlin Heidelberg, ISBN 978-3-540-76938-5, 2008.

T. von Clarmann, N. Glatthor, M. Höpfner, S. Kellmann, R. Ruhnke, G. P. Stiller, H. Fischer, B. Funke, S. Gil-López, and M. López-Puertas. Experimental evidence of perturbed odd hydrogen and chlorine chemistry after the October 2003 solar proton events. *Journal of Geophysical Research (Space Physics)*, 110(9):9–+, August 2005.

J. M. Wallace and P. Hobbs. *Atmospheric Science, An Introductory Survey*. J.M. Wallace and P.V. Hobbs. Academic Press. ISBN 0-12-732951-X, 2006.

R. Wang. Statistical characteristics of solar energetic proton events from January 1997 to June 2005. *Astroparticle Physics*, 26:202–208, October 2006.

R. P. Wayne. *Chemistry of Atmospheres*. third edn. Oxford: Oxford University Press, 2000.

M. Weber. Solar activity during solar cycle 23 monitored by GOME, Proc. European Symposium on Atmospheric Measurements from Space (ESAMS'99), ESTEC, Noordwijk, The Netherlands. *ESAMS'99, Symposium*, 161:611–616, January 1999.

H. Winkler. *The response of middle atmospheric ozone to solar proton events in a changing geomagnetic field*. PhD thesis, AA(University Bremen, Institute of Environmental Physics, Otto Hahn Allee 1 , D-28359 Bremen, Germany), 2007.

H. Winkler, S. Kazeminejad, M. Sinnhuber, and J. Notholt. Conversion of mesospheric HCl into active chlorine during the solar proton event in July 2000 in the northern polar region. (Currently under submission !!!). *Journal of Geophysical Research*, 2008a.

H. Winkler, M. Sinnhuber, J. Notholt, M.-B. Kallenrode, F. Steinhilber, J. Vogt, B. Zieger, K.-H. Glassmeier, and A. Stadelmann. Modeling impacts of geomagnetic field variations on middle atmospheric ozone responses to solar proton events on long timescales. *Journal of Geophysical Research (Atmospheres)*, 113:2302–+, January 2008b. doi: 10.1029/2007JD008574.

J. M. Wissing, J. P. Bornebusch, and M.-B. Kallenrode. Variation of energetic particle precipitation with local magnetic time. *Advances in Space Research*, 41:1274–1278, 2008.

S. Yashiro, N. Gopalswamy, G. Michalek, O. C. St. Cyr, S. P. Plunkett, N. B. Rich, and R. A. Howard. A catalog of white light coronal mass ejections observed by the SOHO spacecraft. *Journal of Geophysical Research (Space Physics)*, 109:7105–+, July 2004.

J. Zhang, K. P. Dere, R. A. Howard, and A. Vourlidas. A Study of the Kinematic Evolution of Coronal Mass Ejections. *The Astrophysical Journal*, 604:420–432, March 2004.

Y. Zhang, W. Sun, X. S. Feng, C. S. Deehr, C. D. Fry, and M. Dryer. Statistical analysis of corotating interaction regions and their geoeffectiveness during solar cycle 23. *Journal of Geophysical Research (Space Physics)*, 113:8106–+, August 2008.

Danksagung

Während meiner Doktorandenzeit an der Universität Bremen konnte ich glücklicherweise mit der Hilfe von vielen Personen rechnen. Bei diesen Personen möchte ich mich nun hier in Kürze bedanken :

Allen voran, Dank an meine Betreuerin Dr. M. Sinnhuber, die mir die Möglichkeit gegeben hat in diesem interessanten Gebiet zu arbeiten und mir mit Rat und Tat zur Seite gestanden hat.

Besten Dank an Prof. J. Notholt für seine Betreuung und Ratschläge in Zeiten des mässigen Erfolges meiner Arbeit.

Weiters möchte ich mich auch bei meinen Kollegen Dr. B. M. Sinnhuber und Dr. C. von Savigny bedanken, die mir mit Ihrem Fachwissen oft eine sehr grosse Hilfe gewesen sind.

Zu grossem Dank verpflichtet bin ich auch Dr. H. Winkler für die erfolgreiche Zusammenarbeit und seine Unterstützung mit Ergebnissen von Modellrechnungen.

Besonders glücklich konnte ich mich schätzen in meinem Büro gleich zwei wunderbare Freunde zu finden, nämlich Dr. M. Scharringhausen und Mag. G. Kiesewetter ("soviel Zeit muss sein"). Ich möchte mich bei euch herzlich für die tolle Zeit, sowie eure wissenschaftliche und freundschaftliche Unterstützung während dieser Jahre bedanken. Es waren schöne Jahre, und ich kann mir keine besseren Bürokollegen vorstellen!

Besten Dank an alle Kollegen am IUP für das produktive und kollegiale Arbeitsklima.

Während meiner Doktorandenzeit erhielt ich natürlich auch sehr viel Unterstützung und Rückhalt aus meinem privaten Umfeld. Unter denjenigen möchte ich mich als erstes bei meinen lieben Eltern Hamid und Judith bedanken, die mich seit beginn meiner Ausbildung nicht nur finanziell sondern auch emotional unterstützt haben und immer hinter mir gestanden haben.

Weiters möchte ich meinen Geschwistern Natascha, Bobby und auch meinem Onkel Andreas dafür danken, dass sie immer für mich da waren und mit mir wunderbare Zeiten gemeinsam durchlebt haben.

Besonders dankbar bin ich im Speziellen den Leuten, die es geschafft haben mir Bremen zu meinem zu Hause zu machen und immer um mich und für mich da waren (aufgelistet in alphabetischer Reihenfolge): Anahita , Andreas, Bobby, Conny, Farid, Gregor, Martina, Matze (Hr. SHK), Marco und Thomas.

Acknowledgment:

The data used in this study were acquired as part of NASA's Earth-Sun System Division and archived and distributed by the Goddard Earth Sciences (GES) Data and Information Services Center (DISC) Distributed Active Archive Center (DAAC). Special thanks also to J.E. Johnson NASA Goddard Space Flight Center, M. McHugh and R.E Thompson from the GATS-Inc. for their information about the error estimation.

Die VDM Verlagsservicegesellschaft sucht für wissenschaftliche Verlage abgeschlossene und herausragende

Dissertationen, Habilitationen, Diplomarbeiten, Master Theses, Magisterarbeiten usw.

für die kostenlose Publikation als Fachbuch.

Sie verfügen über eine Arbeit, die hohen inhaltlichen und formalen Ansprüchen genügt, und haben Interesse an einer honorarvergüteten Publikation?

Dann senden Sie bitte erste Informationen über sich und Ihre Arbeit per Email an *info@vdm-vsg.de*.

Sie erhalten kurzfristig unser Feedback!

VDM Verlagsservicegesellschaft mbH
Dudweiler Landstr. 99
D - 66123 Saarbrücken
www.vdm-vsg.de

Telefon +49 681 3720 174
Fax +49 681 3720 1749

Die VDM Verlagsservicegesellschaft mbH vertritt

Printed by Books on Demand GmbH, Norderstedt / Germany